Inventors at Work

The Minds and Motivation Behind Modern Inventions

Brett Stern

Apress·

Inventors at Work: The Minds and Motivation Behind Modern Inventions

ISBN-13 (pbk): 978-1-4302-4506-3

ISBN-13 (electronic): 978-1-4302-4507-0

Trademarked names may appear in this book. Rather than use a trademark symbol with every occurrence of a trademarked name, we use the names only in an editorial fashion and to the benefit of the trademark owner, with no intention of infringement of the trademark.

Distributed to the book trade worldwide by Springer-Verlag New York, Inc., 233 Spring Street, 6th Floor, New York, NY 10013. Phone 1-800-SPRINGER, fax 201-348-4505, e-mail orders-ny@springer-sbm.com, or visit www.springeronline.com.

For information on translations, please contact us by e-mail at info@apress.com, or visit www.apress.com.

Apress and friends of ED books may be purchased in bulk for academic, corporate, or promotional use. eBook versions and licenses are also available for most titles. For more information, reference our Special Bulk Sales–eBook Licensing web page at www.apress.com/bulk-sales. To place an order, email your request to support@apress.com

The information in this book is distributed on an "as is" basis, without warranty. Although every precaution has been taken in the preparation of this work, neither the author(s) nor Apress shall have any liability to any person or entity with respect to any loss or damage caused or alleged to be caused directly or indirectly by the information contained in this work.

Mom, thanks for getting me all those LEGOS to play with and not making me clean up my room.

I was busy inventing.

Contents

About the Author

Brett Stern is an industrial designer and inventor living in Portland, Oregon. He holds eight utility patents covering surgical instruments, medical implants, and robotic garment-manufacturing systems. He holds trademarks in 34 countries on a line of snack foods that he created. He has worked as an industrial design consultant for such clients as Pfizer, Revlon, and Saatchi & Saatchi, and as a costume materials technologist for Warner Bros.

Stern has been an instructor in industrial and product design at the Art Institute of Portland and Parsons School of Design. He has lectured on innovation and biomedical technology at Columbia University and Stanford University. He graduated from the University of Cincinnati with a degree in industrial design. Stern's work may be viewed at www.brettstern.com. Visit this book's website, www.inventorsatwork.com, and contact the author by email at invent@inventorsatwork.com.

Acknowledgments

Seemed like everyone in Portland, Oregon, was writing a book except me. One day I got a call from Matt Wagner of the Fresh Books literary agency, asking me if I wanted to write this book. I got off my bike, put down my beer, and surrendered to the muse. Though I do know how to invent stuff, but really don't understand where all the commas are supposed to go, the publisher, Apress, teamed me up with several people who really obsess over proper punctuation and grammar to move this project forward:

Jeff Olson, Acquisitions Editor
Robert Hutchinson, Development Editor
Rita Fernando, Coordinating Editor
Kimberly Burton, Copy Editor

I greatly appreciate all of their efforts. They made it fun, and isn't that the whole point?,!,?,!,?

As I've asked my interview subjects about the benefits of having mentors in their lives, I would like to thank the mentors in my life: Michael Cousins, Richard Thompson, and Matthew Lesko.

And, finally, to all the people who said no to my crazy ideas. Thank you for the motivation to keep moving forward.

Introduction

Inventions in steam engines, cotton mills, and iron works converged in the eighteenth century to propel the First Industrial Revolution. Inventions in internal combustion engines, electrification, and steelmaking in the nineteenth century ushered in the Second Industrial Revolution. Today, twentieth-century inventions in digital technology are being conjoined with twenty-first-century innovations in software, materials and advanced manufacturing processes, robotics, and web-based services to inaugurate the Third Industrial Revolution.

Today, at the dawn of the nexus of the future, ideas for inventions stand only a small chance of being realized and competing in the marketplace unless they're generated or picked up by corporations that can marshal teams of scientists and lawyers underwritten by enterprise-scale capital and infrastructure. None-theless, millions of individuals still cherish the dream of inventing and building a better mousetrap, bringing it to market, and being richly rewarded for those efforts. Americans love their pantheon of garage inventors. Thomas Edison, the Wright Brothers, Alexander Graham Bell, Bill Hewlett and Dave Packard, and Steve Wozniak and Steve Jobs are held up as culture heroes, celebrated for their entrepreneurial spirit no less than their inventive genius.

This book is a collection of interviews conducted with individuals who have distinguished themselves in the invention space. Some of the inventors interviewed here have their Aha! moments in government, institutional, or industrial labs; develop their inventions with multidisciplinary teams of experts; and leave the marketing of their inventions to other specialists in the organi-zation. Other inventors in this book develop their inventions with small teams in academic labs and try to translate their research into product via licenses or start-up companies. Still other inventors carry on the classic lone-inventor-in-his-garage tradition and take on the task of bringing their products to market themselves. And a few mix and match their strategies, bringing skills honed in big labs home to their garages and licensing their personal inventions to big corporations.

This is not a recipe book. It doesn't aim to reduce invention to a foolproof sequence of steps: take an idea, go through the R&D process, develop proto-types, create intellectual property, build a brand, raise capital, and get product on store shelves. Rather, this book invites readers to touch their own creative impulses to the fires of real inventors speaking candidly about what possesses them every day of their lives: the passion to question the status quo and invent the future.

These privileged conversations have confirmed my belief that inventors are born to be inventors. Of the multitudes of clever people who get technical degrees in engineering and scientific fields, only a fraction contrive to add patents to their credentials. It is the rare individual who can combine curiosity, intellectual powers, mechanical knack, and focused awareness to see a novel solution to a problem, and then convert that insight into a physical invention that works. (For the sake of coherence, I decided to interview only the inventors of physical products, and not the inventors of software products alone.)

This collection of interviews shows how a startling variety of inventors selected for their widely divergent backgrounds, educations, fields, interests, personalities, ages, gender, ethnicities, business circumstances, and invention types—nonetheless share the ability and indeed the compulsion to create ideas and objects that are useful, exciting, and unprecedented. Many of the inventions described in this book by their creators have transformed our world; others simply made it a more fun place to live. But my intent in these interviews was not primarily to have the inventors talk about the details and significance of their inventions, which could scarcely be touched on in the course of a single conversation. My aim was instead to elicit self-revelations about the invention process and the creative personality.

What I found is that all of these very different inventions sprang from a set of common traits in their inventors: perseverance, drive, motivation, a touch of obsession, and—perhaps most importantly—a buoyant inability to see experimental failure as anything but a useful and stimulating part of the invention process. I was also struck by the fact that most of the inventors I interviewed expressed a similar set of preferences and work habits. They like to work on multiple projects simultaneously in multidisciplinary teams, freely sharing their ideas with others. They reach out to experts in other fields and ask lots of questions. They wake up in the middle of the night and sketch out their ideas on paper or visualize them vividly in their heads. They prototype ideas using materials that they are comfortable working with. They use physical exercise to relax their minds and jack up their concentration. They seek mental stimulation and different tempos of thought in areas outside their specialties. Most strikingly, they value slow time to ponder, dream, and free-associate. So ingrained are these traits and habits of mind that none of the inventors I interviewed could imagine ever ceasing to invent, even if they retired from their professions.

As for any project, I started this book with a clean piece of paper, on which I drew up a dream list of inventors and inventions that especially piqued my curiosity and admiration. And it was literally a dream list, since my underlying motive in undertaking the project was to investigate how dreamers have changed or influenced the everyday waking world of the rest of us.

The breadth of this book from conception to consumption illustrates how profoundly and rapidly our world has been transformed by these dreamers. I sat musing at my desk, then I Googled around on my Mac, and then I recruited my dream team on my iPhone. I conducted the interviews over Skype while recording the conversations using Call Recorder software, which I converted into MP3 files and then uploaded to Dropbox, from which a specialist transcribed them into Microsoft Word documents, probably with the aid of voice-recognition software. Then I uploaded the manuscript to my publisher's interactive platform, on which I collaborated with a team of brilliant editors and production people scattered all around the country, as cozily as though we were all sitting together around a table. You are reading this dialogue either in a paperback book that was printed almost instantly on demand, or on an e-book that you instantly downloaded from the cloud online from a vendor of your choice in whatever format you desired. Perhaps you are at this very moment being moved to tweet or blog about your insights learned from this book. Just imagine all of the inventions and inventors required to fulfill these tasks.

Even if you don't go to such rapturous lengths, I hope that you will join me in my homage to the inventors, engineers, and mad scientists who have reshaped our world. My goal in creating this book has been to inspire you to tap your creativity, to invent something really awesome, and to make the world a better place. Please, please follow your dreams and don't give up.

Gene Frantz

Principal Fellow
Texas Instruments

Gene Frantz always knew he wanted to be an engineer. He spent much of his youth taking things apart and subsequently earning his parents' clemency by convincing them it was all in the name of discovery. Now, as a Texas Instruments principal fellow and one of the industry's foremost experts in digital signal processing (DSP), Frantz continues to bend the rules to propel innovation.

Regarded as "the father of DSP" by many in the industry, Frantz has been intimately involved with the evolution of the technology—from theory, to product, and now to its phase as a true catalyst for new markets and products.

Having joined TI's consumer products division in 1974, Frantz helped lead the development of TI's educational products. He served as program manager for the Speak & Spell learning aid and headed the development team for all of TI's early speech products. Frantz is an Institute of Electrical and Electronics Engineers (IEEE) Fellow, and he holds more than 45 patents in the area of memories, speech, consumer products, and DSP.

Frantz received his BSEE from the University of Central Florida in 1971, his MSEE from Southern Methodist University in 1977, and his MBA from Texas Tech University in 1982.

Brett Stern: So what does a principal fellow do all day?

Gene Frantz: Many years ago, a fella wrote a book on "intrapreneurs," which are those entrepreneurs that actually still have a job in a company. And you find out very quickly that intrapreneurs get fired a lot.

Stern: You're still there, so obviously you've done something either right or wrong.

Frantz: Let me define the term "firing" in these terms: I've been fired six or seven times at TI. And, don't write that down as derogatory yet. I tell people that when two words come up in conversation, I get fired. They are "schedule" and "profit." And the advantage that I have—as a friend of mine described it—is that TI is a start-up company with thirty thousand employees.

Stern: Well it's good that they can continue that frame of mind.

Frantz: But, if you think about having that much resource behind you, there comes a point when the innovation is over and it's time to make money. That's a different skill set. Now this was a different skill set than I had, and so it was easy to "fire" me and find a good business manager who could come in and take over, and run a business rather than a hobby shop.

Stern: So you're never really involved in the moneymaking side of it.

Frantz: No. That's correct. They learned to fire me far before [I got us into trouble]. I am a user of money and as I tell people, I am thankful every day of my life that almost one hundred percent of TI engineers are not innovators.

Stern: What are they then?

Frantz: Good, solid, thriving, development engineers who can make things happen and make them happen over and over and over and over again.

Stern: What is your definition of innovation?

Frantz: I have this little chart that I think came from public radio that says: "Knowledge is knowing the right answer or having the right answer. Intelligence is asking the right question." I take it two steps further. Creativity is asking the question for which there is no answer. And innovation is answering that question. There is another one actually below that: Business, which is making money off of the answer.

Stern: So what is your job?

Frantz: I have been called a serial innovator.

Stern: Your job is to find the answer?

Frantz: Yes. And sometimes ask a question. But generally, the question doesn't give you the start of a business. What gives you the start of the business is the answer to that question.

Stern: Well, in this process—because the process is very important—who asks the questions or who finds the questions?

Frantz: Sometimes I ask the questions. Sometimes the questions are asked by someone that didn't even know they were asking it.

Stern: Is that going into the marketplace, if you will?

Frantz: I spend about a third of my life at universities or small start-ups, listening.

Stern: What about listening or watching the actual end user, the consumer?

Frantz: That too. But I notice that with consumers—if you watch them correctly—you get the answer. But in many cases, the consumer doesn't know what they want until they've seen it.

Stern: Is it that they are afraid of new things or afraid of change?

Frantz: I think a bit of it has to do with that they don't know the capabilities of the technology.

Stern: So all the marketing out there—the market research and all those people spending their days asking for consumer feedback—what does that all mean?

Frantz: Well, you're right. That all works. I'm going to give you an example that I think is fun, which might get you down the path. The University of Southern California is doing research in the area of artificial vision. One of the research teams I glommed onto is creating a camera that fits in the eye, so it didn't have to put on a set of glasses. This camera is about the size of a grain of rice. Now in this case, I asked the question, "Where else can I put this camera?"

We did a lot of brainstorming on that and came to the conclusion that the camera, the size of a grain of rice, could actually create a $10 billion market opportunity. Use as an intraocular camera for artificial vision might be as much as $10 million.

The actual purpose it was designed for was only one-tenth of a percent of its value.

Stern: What were some of the variations you had for the tiny camera?

Frantz: I'm not going to tell you that.

Stern: Well, can you give me some background on how you got here before we really get into the interview? Your background, where you were born, your education, and your field of study?

Frantz: It's a fairly simple education—a bachelor's degree out of the University of Central Florida, and an MSEE out of Southern Methodist, and then an MBA out of Texas Tech. What I find about education is that, in more cases, education kills innovation than encourages it.

Stern: Obviously, you had some interest in inventing growing up. How did that first appear?

Frantz: Oh, I don't know. I had a third-grade teacher that sent a note to my mother that said, "Gene just stares out the window," and my mother sent back a note that said, "He must be thinking great thoughts," and the teacher sent back a note that said, "Nope. He's just staring."

Stern: And what do you think you were thinking?

Frantz: Just staring.

Stern: So you didn't have inventions when you were a kid?

Frantz: No. I just don't remember. But, I think what made it all come together when I started work at Texas Instruments was being able to be in an organization at a time when we were trying to create new consumer products. I was in the calculator division and hooked up with a kid, Larry Brantingham, who was about my age that was really creative in the area of IC-integrated circuit development. I found myself to be very creative in the area of system development because I had the simple idea of asking the question, "Why does it have to be that?"

Stern: You're known as "the father of DSP technology"—digital signal processing. Can you explain that in technical terms? And then can you explain it in layperson's terms?

Frantz: First of all, I'm always careful not to take credit away from the people of theory who created the whole science of signal processing. I was more on the "let's go make it happen and prove it can be done" side. And it wasn't me alone. There were three other guys that worked with me. One of them was that same young kid. He and I worked for the consumer business at the time, and we were the two kids somebody would [come to and] say, "Here's a new idea. Why don't you see if you can do it." And we would figure out how to make it happen.

Stern: So you were the practical arm of the idea.

Frantz: The systems arm of the idea.

Stern: Could you explain the technology in technical terms?

Frantz: DSP, simply put, is based on the idea that all interesting signals in the world are analog. Once I say that, there'll be somebody in some business who says I'm wrong, but the heck with 'em. Sound is analog, vision is analog, feel is analog, taste is analog, and all of those are analog signals. The advancements in integrated circuit technology for decades has been on the digital side rather than the analog side, so if I wish to manipulate or use those analog signals to gain information and to do interesting things, then I need to do them in what is called digital signal processing. For example, your cell phone is fundamentally a digital signal processing solution. Your MP3 player is a digital signal processing solution. Digital TV is obviously a digital signal processing solution. Virtually everything we do now is driven by this concept of digital signal processing. Fundamentally, it is the mathematics that answers the question, "How do I mathematically model a system and then make it work?"

Stern: So your job back in '76 was to take this theory of how to do it and actually make a chip to do it.

Frantz: That was the other team's job—to make a chip. Mine was to make the whole system work. Understand that just because you made something speak—and that's what we were doing—doesn't mean it's a good idea. And if you go back and read the literature from the late seventies and early eighties, you'll find that we tried to make cars talk. That wasn't really exciting. As one person said [at the time], "Everybody knows a door is not a jar. Why does the car say, 'Your door is ajar?'"

And we made elevators talk. We made everything we could think of talk, and in most cases, people said, "I don't want it to talk. Would you please shut it up?" So there was this desire around the world of "I want to put speech on everything," and then a backlash of "but I don't want speech on everything." So it became [a question of], "Well, what things would best work with speech? Which ones would work without it?"

Stern: So where did the application get applied?

Frantz: What I began to do is to work with companies and try to help them understand what speech capability was, what the limitations were, and when they stepped over the bounds of usefulness.

Stern: The marketplace sort of defined where the technology would be applied.

Frantz: Yes, but the marketplace, as usual, doesn't understand when it works correctly and when it doesn't work. One of the products we came out with about a decade after the *Speak & Spell* was the Julie doll. She was a doll that had speech recognition on it.

Stern: And how did that go?

Frantz: Well, there were other things that made it die a short death. It happened to be that they brought it out in late 1987, which if you remember, there was the crash in the stock market, and start-ups didn't do very well through that crash. But I had been working with toy companies for years trying to add speech recognition to their products, and it really came down to this silly notion at that time that speech recognition did not work.

Stern: Did the companies know how the technology worked in those situations? Did the companies come to you looking for something? Or were you going to the companies, saying, "I have a great solution."?

Frantz: A little bit of both. I went after companies, saying, "We have this new technology—now what can you do with it?" And companies came to me saying things like, "We have a brilliant idea, and all we need is your speech recognition capability to make it work."

Stern: So back in the day, what was the prior art? Or what else was going on in the industry with this technology?

Frantz: Well, that particular one was mostly used in military systems to do specific things, of which you spent more time training the user than training the product. There were just too many problems. A lot of it was how you match the problem to the technology.

Stern: Your career has been in a corporate setting. Can you talk about that as far as being an inventor person, and then being part of a team that went out and developed the technology or commercialized it?

Frantz: In a corporate setting in a place like Texas Instruments, we fundamentally make our money selling integrated circuits and selling a lot of them. So we're looking at where we can sell products—areas that could use our devices that we haven't thought about up to this point. I could give many, many examples of working on a new business start [many would be certain] was not an interesting area. Yet as we pursued it and made it possible, it became very, very much a large part of the company.

Stern: So whose responsibility is it to get rid of the old stuff and sort of embrace the new stuff?

Frantz: We don't really get rid of stuff in that sense. The market actually decides to quit buying our old stuff and it goes away. We still have things that sell. TTL [transistor-transistor logic] was introduced in the early seventies. We still sell it today. Many of the early DSPs we created, we still sell today. So old stuff stays around as long as our customers can innovatively figure out how to use them.

Stern: Does Texas Instruments work such that you have a technology and then your group just tries to find new uses for it?

Frantz: Yes. When you find a new market area that is interesting, your first attempt is to build a business based on products that already exist. And then as you get some success out of that, you begin to look at how you could add something to a device that's already in design for another market. Usually it's at the third or fourth generation when you begin to say, "Well, now, let me do an actual integrated circuit specifically for this market."

Stern: Generally, is there a timeline to go from first to fourth generation?

Frantz: I usually assume each generation is somewhere between three and five years.

Stern: Okay, so it's almost a generational lifetime then.

Frantz: Yes, it's quite a long time. And if you think of most corporate situations, at least in integrated circuit technology where three years is a fairly long time, you say, "Well, that's an extremely long time to wait for a market to take off."

And that's why it is best to start that first one with effectively no investment specific to that new market opportunity.

Stern: But, you have all that manufacturing capacity behind you, so that's a pretty big investment.

Frantz: Yes, that's correct. That's back to the start-up with thirty thousand employees.

Stern: Going back to that thirty thousand employees, you talked earlier about this team effort. I think that there is this urban myth of the inventor. A lone guy in a garage, if you will. Can you talk about team effort and how you divide responsibilities?

Frantz: And it's back to something you caught me on earlier, and that is that creative moment of "Aha! I think we can go do this." By the time it gets to the end user, you may have had hundreds of people doing innovation to get it there.

Stern: The "aha" moment is a fraction of a second.

Frantz: Yeah, yeah, yeah. The creative movement is really an "aha" moment. And it's a lot of hard work thereon.

Stern: So basically the "aha" moment, which is the fun time, is a fraction of a second, and then you really have to just sit down and work after that.

Frantz: There are a lot of issues that have to be solved. I tell engineers and particularly engineering students that they have spent their four years at a university learning how to solve problems. Do they ever worry that we're running out of problems? And then I inform them that that's my job I create problems.

Stern: So, you are a problem creator. Well, in the team, whether you're managing the team or being part of the team, how does it get divided as far as the skill sets or the tasks for each to find the problem? You put three engineers in a room and you say, "Solve this problem." How does it get figured out?

Frantz: First off, you don't put three engineers in a room randomly. If you look at that team of four that started the *Speak & Spell*, we each had complementary skills and we were also at a point where we had nothing to do. And by the way, the creative "aha" moment was by my boss, not the other three of us.

Stern: If there are four people in a room, all bright individuals, and you get several directions to go down, several paths if you will, how do you decide which is the correct direction?

Frantz: When I worked through my MBA, one of the lessons that I learned from my business class was there are two reasons a start-up goes out of business. One is they have only one idea and the other is they have too many ideas.

Stern: So what's the right number?

Frantz: The right number is many ideas, but you focus on one until it's completed and then you go to the next.

Stern: In talking about the idea, can you explain or define your ideation process then?

Frantz: When I talk about innovation, I talk about how there are two ways of innovating. One is to create technology, and then figure out where to use it. And the other one is to figure out an itch that hasn't been scratched yet, and figure out what technology you have to pull through to make that happen.

And I go for the latter, so I am busily looking for that new idea, that new thing that nobody knows they need yet, but once it's available will say, "Well of course I have to have that."

Stern: You focus on finding the problem.

Frantz: Yes.

Stern: As compared to finding the solution?

Frantz: That's right, because the solution, in my view, is fairly straightforward to define and determine whether you have an opportunity to make it viable. I'll chase a rabbit here for just a second. I tell researchers that if they want to figure out new research topics to pursue, to go back and read the papers from twenty or thirty years ago to find out all those areas where people said, "This is really, really neat, but the technology is not available to do it." It's probably available to do it today.

Stern: So what you're saying is there's generally a twenty-year time lag between solution and implementation?

Frantz: There could be, and that's part of "Aha! I have this new idea." Is the technology available today? Will it be available in three years? Could it be available in three years if we pushed hard?" In other words, if you say naturally it will be ten years before we get there, can I pull it in [closer to today] by being creative?

Stern: After you see this problem or the possibility of a problem, do you write it down? How do you really define what that problem is?

Frantz: I write it down. A lot of times, I'll do a presentation because I need pictures to show this. I tell people the appropriate way to do a presentation is with a big font, short words, and lots of pictures—just because that helps get the base information down and allows for a lot of creativity to be filled in.

Stern: Are the pictures of situations or actual objects?

Frantz: Could be objects, could be drawings, could be whatever. It still works out that in many cases it is on the back of a napkin.

Stern: So you have this graphic representation of the idea. What are the next steps to ideate or brainstorm? Do you sketch? Do you prototype? What is the process that you personally go through or some examples that your team members go through to get to that solution?

Frantz: I do a lot of brainstorming with people.

Stern: So just sitting around a room and talking.

Frantz: Yeah. Or going to the whiteboard and screaming and yelling at each other. Or talking to an audience and getting them to give feedback. A lot of what I'm doing—after having this brilliant idea and keeping it to myself and figuring out how to make money, in many cases, since it is the end equipment that I am innovating and TI as a corporation does not sell end equipment but the components that go into it—I spend a lot of time throwing ideas, half-baked ideas, out to our customers and letting them finish the idea.

Stern: There is so much negativity toward new thoughts, new thinking, and new ideas, how do you personally have that confidence to say something that is off-the-wall and half-baked? What gives you the ability to do that?

Frantz: I've done it my entire career, so everyone's kind of used to these crazy ideas.

Stern: What do you think prevents people from having that ability?

Frantz: I have lots of people who will say to me, "You can do that because of the position you hold."

Stern: Which is true.

Frantz: And I say the reason I have this position is because I've been doing this my entire career. I tell people that innovation is the sport of young people. And there are a couple reasons I say that. One is that it is a twenty-four-hour, seven-days-a-week activity to innovate and create a new product. There is a second reason, and that is people with reputations are more interested in protecting their reputations than taking risk. And young people have no reputation to lose.

Stern: So you feel just by the nature of youthful indiscretion that you have that ability to throw ideas out there and say, "What if . . . ?"

Frantz: Yes. And as I say, it is a sport of young people, but you need to have gray-hair types around to keep them pointed.

Stern: How do you control the failures?

Frantz: I don't think you ever control failures. You just realize they are a failure and go on to the next one.

Stern: How do you recognize failures?

Frantz: There are two types of failures. First, there are technical failures, where the technology actually didn't work. Second, the technology was successful but there was no market for it, no customer. And so if I go back to that Julie doll, for example, it was a marketing failure. But it was a brilliant piece of technology and I say brilliant because there were several people that caught the idea and the sense of what had to happen, and did wonderful design to make it work. And I was the instigator and not the guy with all those brilliant ideas.

Stern: Where do you see how the technology—transforming analog to digital information—has been most influential in society?

Frantz: I'd say that it's hard to specify one area.

Stern: Well, how about several areas then?

Frantz: The cloud. That's all a result of this thing called digital signal processing. Now you'll hear somebody else say, "No, it was actually the computer," but those pipes getting your information to the cloud are all DSP-based. All the gathering of the information and putting it into a form of data and then taking the data and making different information out of it—that's signal processing.

If you look at the way we handle our music, the record has gone by the wayside, the CD is almost gone, and now everything we do is either on a computer or on a memory stick. If you look at our automobiles, many of the safety issues lately have been the whole concept of how do I create braking systems that work better than the human can actually consider. That's a signal processing task. If you look at airbags, that's a signal processing task. If you look at cars now that can parallel park themselves, that's a signal processing task. If you look at adaptive driving or adaptive cruise control, that's a signal processing task. You just keep going down the line and you find out there aren't many aspects of our lives that have not been affected by this silly thing that thirty years ago we thought was impossible—and that's signal processing.

Stern: In the next thirty years, where do you think the technology will be?

Frantz: It's easy to put on an evolutionary hat and say it will be smaller, easier to use, implantable, and more invasive in our lives. I think there's a whole set of "aha" moments that we haven't thought of that will occur, which we cannot predict. We can only look back and say, "Well yeah, obviously."

Stern: You just said that technology would be invasive in our lives. That, in a sense, has somewhat of a negative connotation.

Frantz: Of course.

Stern: How do you balance the positive and the negative?

Frantz: That's a hard thing to do. And the reason I say it's a hard thing to do is because as an engineer and as an innovator, my job is to create the capability, not necessarily to decide on the morality of its use. And that always irritates me when I talk to engineering students. I tell them that they are at the beginning of some of the greatest opportunities to make an impact on society—so please make sure it's a positive impact.

Stern: And how do you control that moral judgment?

Frantz: I'll just basically say that's not an engineering activity. That's a community activity. But let me go back to an example. One of the things we added to a lot of our consumer products many, many years ago was text-to-speech. Many of the text-to-speech systems at that time would not allow you to spell a dirty word and have it pronounced. We chose that if you were going to spell a dirty word or an obscene word, [the product] would pronounce it correctly. And we did that because we felt it was our job to be true to the science and not be the moral compass of the user.

In fact, on the *Speak & Spell* program, I remember an instance where we had a parent call in concern that her little Johnny was typing dirty words into the *Speak & Spell*. Why didn't we stop that from happening? And my response was, "Well, first of all, did he spell them correctly? Because that is the purpose of the product." And secondly—and here's why they wouldn't let me talk to the parent—"I'd like to have a discussion with you on parental guidance." It was not our job to be the moral compass for that child. It was the parents' job.

You know, any time you create a new technology, you know that it is going to be misused as well as used properly. If you decide not to offer technology because it might be misused, you lose the opportunity for the good it can provide. I was on a panel many years ago at a major conference, and this very topic came up. I made a comment that it was frustrating to me that in many cases the early adopters of a new technology were either pornographers or con artists. Well, the rest of it was an interesting debate between an editor of a magazine and me about whether that was appropriate or not. But it really comes down to this it does not stop me from considering that next innovation.

Stern: Where do you seek and where do you find inspiration or solutions?

Frantz: I told you I was a serial innovator. Most of it is listening to people, and then taking a different view of what they said or what they were thinking, and describing our product.

Stern: Going back to when you were in the third grade staring out the window—I'm talking about the inspiration. Where do you come up with a possible solution for things?

Frantz: Oh, my engineering background tells me what's possible. I have the background to have a pretty good feel of what I can do, or what we can do and what we cannot do. Then it's just a matter of that "aha" moment—"Well then, why don't we do this?"

But let me just talk you through a story. When I give a talk on cloud computing, I show a slide of a cloud with clutter around it and I always say it's clutter because it's disorganized and it should be disorganized. One of the things I show around that cloud is a cup. Then I talk about what I think would be a great product: I go into Starbucks wanting a cup of coffee so I can sit down at a table and read my e-mails and surf the web. And I want to spend four, five, six hours there. Well, the trouble is that my coffee is going to last me about thirty minutes. Then I'm going to have to get up and get another cup of coffee and watch my table to make sure nobody steals my computer or any of my stuff.

Starbucks should have this new offer of the infinite cup. I go in and I order the infinite cup. I take it to my table. I sit there. And when the cup is almost empty, it tells the cloud in the Starbucks shop that my cup is empty and for them to come over and refill it with the appropriate coffee—and by the way, charge my credit card a little bit of money. So there's a service that doesn't exist anywhere in the world.

Would that be an interesting service? Yes. Could Starbucks make extra money on it? Yes. In fact, if they sold me a special cup and I came in every Friday to spend three or four hours, it could be set up so as I walked in the front door, my cup would say to the cash register through their cloud, "Gene just showed up. Say hi to him, tell him to go sit over at that table, and you'll bring him his coffee shortly."

Stern: And they'll know if you want cream and sugar.

Frantz: And they might even know that I want my first cup black and hot, I want a cappuccino for the second cup, I want iced coffee for the third cup, and just bring me a glass of water for the fourth cup. So I have a preset menu that they all know about. Now you say, "Well, would that be valuable to Starbucks?" Yes. They could make more money and their customers would be happier.

But then I say, let me spread that out [into other areas]. What if I went to a restaurant and all of the glasses on my table were tied to their cloud? So when my glass of water was nearly empty, they'd come over and fill it without me having to wave them down. Would that be valuable? Yes. I would be a happier customer. I would give bigger tips. So to me there's a cloud, there's an opportunity and, by golly, here's something I could do with it.

Stern: Does that leave out the opportunity for any spontaneity?

Frantz: Spontaneity in what sense?

Stern: Well, everything is predicted beforehand.

Frantz: Oh, you can always add spontaneity to anything, even if it's predictable. I could go in and say I don't want that. I want to change it this time. It's all about customer service, but that's an example of a product that just kind of hit me that might be interesting given every store in the world now has a cloud and I might be able to tie to that cloud. Now, that's a simple "aha" moment. Unfortunately, the technology to make that happen is not such an "aha" moment.

Stern: What has to come about for that technology to catch up?

Frantz: Oh, how do I have a cup of coffee that can actually talk to the cloud? What's my sensor? What's my power source?

Stern: So is that a third party or—

Frantz: Or, it's a customer of mine. I tell stories like this to audiences waiting for somebody in the audience to have the "aha" moment of "Oh yeah. I could go do that and I could make a fortune at it. And I'll use TI components to make that happen." Or they jump to that next conclusion of "Well, wait a minute. Gene just talked about an inanimate object that, tied to the cloud, could be useful. What other inanimate objects in my life could I tie to the cloud and make useful?" And so I tell people an interesting brainstorming session is pick any inanimate object in your room and brainstorm what you could do with it if it were tied to the cloud.

Stern: It seems that you're at a point in your career where you get to mentor. To speak to people and give them direction. Do you have mentors in your life and where do you find them?

Frantz: I probably don't have mentors anymore. You get to a point where the mentors you would like to have had have either died, or quit, or gone away. So no, I think at some point you run out of mentors that are useful to you. And I'm not certain I would say what I'm doing to these younger engineers is mentoring them as much as I am enticing them to think about something new.

Stern: Would you say that you have professional heroes or had professional heroes?

Frantz: Yes, I've had some. Inside of TI, I think you kind of have to point to Jack Kilby as a hero.

Stern: And why's that?

Frantz: He invented the integrated circuit.

Stern: And do you have any inventions or inventors outside of TI? Inventions in your daily life that you really like?

Frantz: I can't really think of any. And I don't mean to sound tongue-in-cheek on that because my strength you might say is that I don't really think a lot. Some people read and learn, but I just kind of listen, and then things come to mind. You can almost say I'm a loner and that's fine.

Stern: Outside the digital world, do you have any products that you like having on your desk or in your home that give you comfort in any way?

Frantz: Not really. How's that? Although if you talk to my wife, she'd tell you I do really crazy things. I have a home theater. The house we bought about ten years ago, my goal was to make sure I had a room for a home theater, and so I have a home theater. I probably use it once every six months, but I'm happy to have it there.

Stern: Are there any technologies that you find not useful that are out there?

Frantz: Of course. I just can't think what they are.

Stern: When you are at a dinner party and you're sitting next to a new person, what do you say that you do?

Frantz: I tell them I'm an engineer at TI and I try to figure out what's going to be capable within five to ten years. And then change the subject to something else.

Stern: Do you have any advice for would-be inventors about the required skill sets you think they need to have?

Frantz: I tell people that it's okay to be crazy, it's okay to have stupid ideas, and it's okay to talk about them because sometimes a stupid idea is the nub of what will become a growing idea. You just don't know when that is. When I talk about the cloud, I always say had the pet rock been popular today, it would be tied to the cloud. I don't know why, but it would be tied to the cloud. And the reason I use that example, if you remember when the pet rock came out, what you kinda had to say was, "Really? Somebody would actually spend money to put a rock on his desk and give it a name?"

Stern: But it was an idea someone had, and he told a good story around it.

Frantz: It was a really stupid idea. They made a lot of money and I don't want to pick on the pet rock particularly, but I could probably go down a list of forty or fifty products that were really, really dumb ideas that people made a fortune on.

Stern: What does that say about the marketplace?

Frantz: That we're tolerant of a lot of interesting ideas.

Stern: That's true. With all these interesting ideas out there, could you talk a little bit about the intellectual property side? What's your method? You have forty, fifty patents out there. What is your responsibility when you're thinking of new ideas? And do you have any comments about the USPTO?

Frantz: I'll leave the US Patent Office alone. I just don't have any negative thing to say about them. They're doing their job the best they know how to do and this is a difficult, difficult area—to try to figure out how to capture intellectual property. What I tend to do—for example on this idea of a cup tied to the cloud—before I went out to make those presentations to the industry to give examples of what you might do with the cloud, I turned in a patent disclosure to TI for us to get that protected. So generally, it's like anything else: you turn in the patent disclosure. My job is to disclose I had an idea, not to determine its value.

Sometimes TI looks at it and says, "That's a dumb idea and we're not going to patent it." That's fine. About half of those patents in my portfolio have to do with a really dumb idea we had in the mid eighties trying to sell DSPs into the TV world.

About four or five years later, I had a patent lawyer call me and say, "You know that patent you had four or five years ago?" I thought we were going to have to restruggle through it again because it was a real pain getting it patented, and he said "No. That's the earliest description I can find of a synchronous DRAM."

Stern: So it had value.

Frantz: Long before there was a synchronous DRAM, we invented it.

Stern: Right. But did you know you invented it back then?

Frantz: Oh heavens. Heck no. It just wasn't called a synchronous DRAM at the time. We knew that we invented it, but it wasn't needed in the industry for another four or five years, and then once it was needed, everybody glommed onto it—and by golly, it did exactly what we had done.

Stern: Just as an overview, do you have any particular final advice for inventors out there?

Frantz: No. Invent. It's a fun thing. And advice to the non-inventors: find out what you're good at and do it. Not everybody can invent, not everybody can create, and what makes money is not necessarily the creation or the invention but the day-to-day work—making today's production just like yesterday's production and selling it into the market.

Stern: What do you do for fun or distraction?

Frantz: Oh, actually I do a real strange thing. I collect baseball cards.

Stern: Any particular year or field?

Frantz: Old ones.

Stern: What's your favorite card?

Frantz: I'm a Yankee fan, so it's my Mantle rookie.

Stern: What card don't you have that you're looking for?

Frantz: It just sold in an auction last week. It's called the Honus Wagner T206 card.

Stern: How much did it go for?

Frantz: $1.2 million.

Stern: And you weren't bidding on it?

Frantz: I'm still married.

Stern: Do you plan to retire at any point?

Frantz: There'll come a time where as a corporate innovator I will outlive my usefulness and I will go be a mentor to small corporations starting up, and I will help them grow into big companies.

Stern: You're going to continue the effort, but on a different scale.

Frantz: Yes.

Stern: Any final words of wisdom you want to offer the marketplace?

Frantz: Not really. I think the crazies out there that are the inventors of the world know who they are, but just won't admit it, and that's pretty typical. I just encourage them to continue to be crazy. I used to call them the lunatic fringe, but that got me in trouble.

Eric Fossum

Image Sensor Physicist, Professor
Dartmouth

Dr. Eric R. Fossum *is a solid-state image sensor device physicist and engineer. He is the primary inventor of the modern CMOS active pixel image sensor used in nearly all camera phones and web cameras, many DSLRs, high-speed motion capture cameras, automotive cameras, dental X-ray cameras, and swallowable pill cameras.*

Born and raised in Connecticut, he received his BS in physics and engineering from Trinity College in Hartford and his PhD in engineering from Yale. In 1990, Dr. Fossum joined the NASA Jet Propulsion Laboratory (JPL) at the California Institute of Technology. He managed JPL's image sensor and infrared focal-plane technology research and advanced development. At JPL he invented the CMOS active pixel sensor (APS) camera-on-a-chip technology and led its development and the subsequent transfer of the technology to US industry. In 1995 he co-founded Photobit Corporation to commercialize the technology.

In late 2001, with over 100 employees and revenue exceeding $20 million per year, Photobit was acquired by Micron Technology, Inc. In 2010, he joined the faculty of the Thayer School of Engineering at Dartmouth as a research professor teaching and performing research in advanced imaging devices.

Dr. Fossum has more than 140 US patents and is a Fellow member of the IEEE. In 1996, he was inducted into the Space Technology Hall of Fame. In 2010, he was named "Inventor of the Year" by the New York Intellectual Property Law Association (NYIPLA). In 2011, he was inducted into the National Inventors Hall of Fame.

Brett Stern: Can you tell me about your background—where were you born, education, and your field of study?

Eric Fossum: I was born in Connecticut and went to a public high school there. I went to Trinity College in Hartford and studied both physics and engineering. From there I went to Yale and worked on my PhD. My field is solid-state devices.

Stern: When you were growing up, would you consider yourself an inventive kid? Were you playing around making things or fixing things?

Fossum: Strangely, not as much as I would have liked. I actually attended a special program on Saturdays at the Talcott Mountain Science Center in Connecticut. There were students always coming up with these neat ideas of things to study and I was very frustrated because I felt like I really couldn't think of anything original at all.

Stern: What eventually got you out of that frustration?

Fossum: It's hard to say. I think it's because at that time I was busy trying to find a problem. I couldn't think of a good problem to solve and when I started finding problems to solve in the course of doing my graduate work, it was very easy to come up with creative solutions. So it was really that old adage, "necessity is the mother of invention." Once there was a clear problem to solve, then it became quite easy.

Stern: I was reading in some of your background information that your great-great-great-grandfather, Benjamin Franklin Johnson, was a steam engineer, and various members of your family have backgrounds as machinists or engineers. Would you say any of that was an influence to you?

Fossum: Well, I didn't know most of those relatives because they were so far back. My father was a mechanical engineer and a creative guy, and so I suppose he must have had a big influence on me. As a teenager it didn't quite seem to be such a big influence at the time. It was more like trying to figure out how not to be like him.

Stern: So it's part of your DNA, but not necessarily in the front.

Fossum: Right. And my brother is also a mechanical engineer, so I guess there must be something to it.

Stern: Could you provide some background—first in technical terms, and then in layperson terms—about the technology and the field of study you are known for?

Fossum: My main area is microelectronics and microelectronic devices. My efforts are in the area of image sensors and the chips that convert light into electronic signals. They are used in cameras and camcorders, and that kind of thing.

Stern: At what point did you go into this direction?

Fossum: I was always interested in what is called artificial intelligence, in kind of a computer science sense, as I was growing up. I had a lot of exposure to computers early on, but then as I became more interested in physics and solid-state devices, I wasn't sure how those two interests connected. But I spent a summer at the Hughes Aircraft Company working on infrared sensors for various applications, and that really sparked my interest in smart imaging. How can we make a smart eyeball? That was really what fueled my interest in image sensors.

Stern: What was the state of the art prior to when you started the investigation?

Fossum: The state of the art image sensor at that time in the 1980s was the "charge coupled device" or CCD. Most were coming out of Japan for use in camcorders for consumer use, and it used to be that a camcorder would run for about an hour with a battery the size of a brick. A large part of that power consumption had to do with all the electronics required to make a CCD operate.

The CCD was a fairly power-hungry device and not so miniature. Of course, it was quite miniature compared to a vacuum tube, which was the previous television camera technology before the CCD. Definitely not something that would fit in your shirt pocket like a cell phone does today. I was at Columbia University after Yale, as a professor for six years. We were working on very high-speed image sensors for very fast cameras and signal processing. We were also working on smart image sensors, the sensor chip that had some smarts built into it, but it was still all CCD-based, which was the prevalent technology at the time.

Stern: Was this work for pure research or was there some industry collaboration going on?

Fossum: I would say it was more applied research but not much industry collaboration at that time.

Stern: Was there necessarily a problem you were solving or was this just sort of a "gee whiz, can we do this?" situation?

Fossum: I would say it was more in the "gee whiz, can we do this?" category as I look back at it. We were trying to solve the general problem of how you can put some of the smarts or some of the computing requirements for vision into the imaging chip itself, but there was no consumer application at that time.

I was generally thinking that it would be good for robotics, but in 1990 I left Columbia to go to the Jet Propulsion Lab, which was part of Caltech. When I came to JPL, I was asked to help solve a problem they were having with CCD cameras that were flying on interplanetary spacecraft. These cameras were relatively large and they consumed a lot of power, and they also were very susceptible to radiation effects in space. I was asked to try to help them improve that situation for the future. And that was the necessity that was the mother of this invention.

Stern: Today, the technology you developed is used pretty much in every single camera phone, correct?

Fossum: That is correct.

Stern: At that time—this is fifteen to twenty years ago—did you have any insight that this was where the technology would eventually be driven?

Fossum: At first I was focused on the problem at hand, which was not even of this earth—it was for space. But after working on it for a while, I realized that this miniature camera technology was not only useful for space but had a lot of terrestrial uses as well.

At the time, the cell phone didn't really exist. I think I used the term "portable videophones" in a 1995 interview. Of course, cell phones kind of grew up after that. And even when the cell phone application was seriously proposed to us after we had left JPL and started a company to commercialize the technology, I didn't necessarily believe in it—that it was the killer application for CMOS[1] image sensor technology.

We were talking to customers that were saying it's a way for users—at that time Japanese teenage girls—to share visual information with their friends when they're shopping or that sort of thing. It sounded fun, but not very important from a commercial point of view.

Stern: So you're in this academic setting. What was the motivation to do tech transfer into the commercial world?

Fossum: We started recognizing that it had a lot of consumer commercial applications, and also at that time NASA shifted its technology mission slightly to say we need to transfer all this cool technology that we're making in the various NASA research labs to US industry and help strengthen US industry.

I spent quite a while traveling around to various companies like Bell Labs, Kodak, and National Semiconductor to promote this technology and urge them to take a look at it because it was a way that the US could recapture that imaging technology, which was now almost one hundred percent in Japan. But, I was a little bit disappointed and frustrated by how slow it was taking these companies to wake up to the opportunity. They weren't moving very fast compared to what I considered to be a real window of opportunity. At one point, we just decided that we should start to fulfill the needs that we saw by starting our own company.

Stern: You're saying "we." Was this work part of a team? And if so, can you talk about the collaborative effort?

---

[1] Complementary metal-oxide semiconductor (CMOS) is a technology for constructing integrated circuits.

Fossum: Well, of course it was a team effort. I was the primary inventor of the technology and from an organizational point of view, the manager of the team. The different team members had different roles, but one of the primary team members happened to be my wife at that time, Dr. Sabrina Kemeny. As it turned out, when my youngest daughter was born and Sabrina was home on maternity leave, it was just right as this whole opportunity was blossoming. And so we decided to get her a little computer CAD system at home and she could start working on some designs in between taking care of the babies. It didn't take long actually for that effort to grow quickly. Pretty soon I was leaving for work in the morning to go to my regular job at JPL, and a couple people were arriving at our house to do the work, and then they would leave and I would come home. I have no idea what the neighbors thought of all that.

Stern: At what point did you decide to leave JPL and then start your own company to do this tech transfer?

Fossum: The company was started at that time when Sabrina was working at home. But, after a while it became pretty apparent that if we were going to make this company a success, we really needed all hands on deck and it was time to take leave from my work at JPL and participate on a full-time basis with that company.

Stern: You said that one of the missions of NASA was to do this tech transfer to industry. Was there an incentive for you to leave and do the start-up, or were they not happy about it?

Fossum: I'd say that's an interesting question. Caltech and JPL are very careful about conflicts of interest. We were very open with them about what we were doing and very careful that there were no perceived conflicts of interest between what I was doing at JPL and what we were doing with the company after hours. For me it was after hours—everyone else was working there already.

Early on, even the concept of licensing the technology to the actual inventors was kind of a, believe it or not, a novel idea for Caltech. But, they decided to also take that leap, maybe a little bit behind other universities that were already doing that sort of thing. But, they took that leap and actually wound up with an equity stake in the start-up company as part of the licensing arrangement. In that part, I would say, we had the full blessing of Caltech and JPL, but I don't think they were happy to see me leave JPL. I think everybody understood what the opportunity was and why I needed to take it.

Stern: Was this a self-funded project to launch, or was this at the beginning of the venture capital scene? How did you get the project moving forward?

Fossum: Yes, good question. Actually, at the beginning, we were doing custom designs for different companies that had very specific needs that could not be met by the CCD technology that was out there. Then we also got assistance from various government organizations through the Small Business Innovative Research program [SBIR]. So we were quite self-funded for a couple of years. But, we decided that some of the requirements we were seeing from various customers were getting to be quite similar and that maybe what we called a "catalog product" would solve many customers' needs. We entered the product business and at that point we needed more capital to continue, and we actually were able to secure funding from strategic partners rather than venture capitalists.

Stern: Were these applications industrial, military, governmental, or consumer at this point?

Fossum: Very few were consumer at that time, even in the late nineties. The biggest consumer item was a web camera. So for example, Intel was an early partner with us in consumer products, and we integrated our technology into their web camera. But a lot of them were also industrial. There was for example, Basler, which is a German company that did inspection systems, and there was Schick Technologies out of New York, which produced dental X-ray systems. They were one of the first companies to make a product where you put a chip in your mouth and the dentist would take an instant X-ray rather than using film. There was also Gentex in Michigan that was working on automotive applications with these kinds of sensors.

Stern: So were you selling a part or were you licensing the technology at this point?

Fossum: We didn't really do any licensing. In the case of a custom design, we would design the part. Some customers would take over the production at some point. But for other customers, we would actually provide a niche product to them. For example, that famous pill camera—where you swallow the camera and it goes down through your intestinal tract—came from Israel, a company called Given Imaging. We would actually sell them chips. We manufactured that chip in a fabless semiconductor model and they would buy parts from us. And then for the web cam kind of application, we would sell parts to companies like Intel or Logitech.

Stern: It seems that some of the ideas are coming from within and then some of the problems come from outside, from industry asking you to provide solutions. Can you explain the difference between those?

Fossum: The basic technology underlying the chips at that time and continuing today was the invention that we made while we were at Jet Propulsion Lab. But, then there are continuous improvements to be made both in the speed, the accuracy, and the image quality of those image sensor chips and the improvements are also inventions in their own right.

Then there are the kind of application problems like Gentex was interested in—like putting a camera in a car, mounted on the rear-view mirror, that would look at oncoming headlights to decide if you should dim your high beams or not. This is now a product, which I believe is in a lot of cars, called SmartBeam. At that time it was a unique problem that they posed for a product that they wanted to make, and image sensor technology was part of the problem. We had to work together as a team with that company to produce a total solution.

Stern: Could you explain the ideation process that you go through once you see a problem?

Fossum: It is so hard to even describe it as a process. I think as engineers we would like to assume that there is a rational process flow for every problem/solution. In my experience, that is not true. Solutions or approaches just kind of flash into your brain, and sometimes after you think about them for a while, you realize maybe it is not such a good idea after all. Other times, you refine that idea and it becomes quite workable. Then there are a lot of details to work out. I was pretty good at coming up with a general solution or strategy or approach. It really was the entire team many times that made that idea work out.

Stern: Do you feel that as an inventor, or an inventive person, you need any particular skill sets to perform these functions?

Fossum: Oh, sure. There are a lot of skills, of course. In my case, you really have to understand device physics and imaging quite well to flash in on solutions that are going to give high-quality imaging performance. You have to understand what has been done, what can be done, and what needs to be done to come up with a solution that is constrained by all those practicalities of what you can do in a semiconductor device or in a circuit, for that matter.

I remember we were given a problem once to try to develop one of the first high-definition image sensors in the CMOS image sensor technology. At the time, it seemed quite daunting to build an image sensor that could produce one or two million pixels at thirty or sixty frames per second. That would be like a thirty or sixty megapixels per second data rate. Within a couple of years, that was kind of routine and we were producing sensors that produced a billion pixels per second in data rate.

Stern: What do you think is the motivation when you see this big wall in front of you? What gets you up in the morning so you say, "Yes, I'm going to tackle a problem that seems insurmountable."?

Fossum: I suppose that is a personality thing. It depends on your personality. I happen to like challenges like that, where I have the skill set at least to be in a good position to attack the problem. I don't know, I just gnaw on a problem 24/7 until I feel like I've got a solution that is going to work.

Stern: When you have this problem in front of you and you're trying to brainstorm in your head or ideate, do you have any methods that you go through? Do you sketch? Do you prototype? Do you get on the computer? What is your process that you use to investigate these various directions?

Fossum: I think you don't get on a computer or prototype until you've got an idea that is well in hand. So the idea just kind of starts in your head and soon has a solution. Maybe I put pencil to paper and try to sketch it out in a little more detail and do a couple of rough calculations to see if it is going to fit the requirements. Then you can let the engineering work really begin. That's when you start using computers for solutions and modeling. At some point, you decide you are going to invest in actually making the part to prototype it, but since that is a fairly large financial investment, you don't take that step lightly.

Stern: In doing all this work for the past twenty years, have there been any really big failures?

Fossum: I won't say that there's been any really big failures. We did do one or two jobs for different customers where the engineering process failed for one reason or another, and the chip that we produced the first time for the customer didn't work very well. That's a pretty intense time for a small company when you've invested a lot of money in a chip and you've got a customer that is really depending upon you to deliver part of a total product that they are trying to make—and your part doesn't work.

I would say that, fortunately, a lot of the skills I acquired because of my work at JPL—working in that space technology environment where you must make things work and understand the engineering process to make things work right the first time—helped us avoid a lot of pitfalls. No one has ever asked me that question before, but I really think it is an important part of the success.

Stern: Did you learn anything after that opportunity?

Fossum: Yes. There was a lot of pain involved. I think the learning was just to redouble our efforts to improve our engineering processes so that it wouldn't happen again. Image sensors are fussy devices to make, and it doesn't take much for things to go wrong quickly. Again, modeling up front and having some careful engineering methodology is important to taking an idea and actually making it work in a product setting at the end. There is a creative step, but then there is a very disciplined engineering step that is still required.

Stern: You've been involved in numerous start-ups. It sounds like you have been the lead person on the commercialization business side. Can you talk about the difference between sitting behind the computer figuring out the technical stuff, while at the same time the business commercialization side?

Fossum: I really enjoy the business side as well. There is a lot of creativity in business as you're trying to find a solution that is win-win for both parties. Creating a contract that makes it very plain as to who is responsible for what and who is going to deliver what, I certainly enjoyed. You are also trying to put the right people in the right place at the right time in the business environment.

You are also trying to make sure that your team is assembled with the right set of people and the right skills.

Stern: Your background and your degrees are certainly on the technical sides. Where did you learn the business side?

Fossum: When I was in my early teens, I would say. I had a paper route. I had to get up and deliver papers every day, and practically everything I needed to know about business I learned from having that paper route.

Stern: Could you give me a few things that you learned?

Fossum: Well actually, it may sound silly but it really is kind of true because the paper company would sell you newspapers, but you would have to go out and sell customers that were in your neighborhood on that product. A large part of that is not only selling the product that the newspaper company is putting out, but also a lot of it is very service oriented. You have to bend over backward to make sure that you are giving them what they need and what they expect.

And there is the financial side, which means that you also have to be able to collect money from the customers and you have to maintain a relationship with them. Of course, then you pay for your product, which is paying the newspaper company every week whatever you owed them. Not to mention the hard work of getting up every morning at five o'clock and delivering papers in the middle of winter, which is not so much fun either. So there was that part of it.

We had various people come in from time to time at Photobit to help educate us. I learned a lot about the concept of selling a whole product to people, not just the chip. They don't need just the chip, they need documentation, they need examples on how to use the chip, they need product engineers that visit them regularly and help them solve their engineering problems in using the chip. This kind of whole product strategy is something I learned from consultants that we hired. In fact, probably our company would have gone under if we hadn't learned that important lifesaving approach to what is called the whole product, which is addressed by some well-known people in marketing. It was news to me as an engineer, but probably not news to people that were already well experienced in high-tech marketing.

Stern: On the technical side, you have more than one hundred forty patents. Could you talk about the intellectual property side, and your motivation and any thoughts on the process in the United States?

Fossum: A patent is a fairly expensive thing, typically. I think my general view is that a patent is basically a way to participate in business with some protection, especially as a new company. It's not really a very good fence for keeping your competitors from using that technology, but it at least allows you to get into the game. So it's really a necessity for a small company to file a patent on key things that they have. There is the strategy of once you try to put in a patent, how many different things you put into it. You pay per application. There are also issues about where you file the patent outside of the United States, if anywhere, and there are a lot of different views on that strategy. Since a lot of my patents are still active. I think it's best to not comment on that thought process.

Stern: When you're thinking of new ideas, when you have a problem in front of you, where do you seek inspiration, or where do you seek solutions? Are they just always based upon past technologies, or do you go out there in the marketplace, or go out in nature—where do you get them?

Fossum: You know, as I was saying before, solutions kind of flash into your brain at some point. I can't really explain what my subconscious activity is. Especially these days, I like to spend a lot of time on my farm pushing rocks with my tractor, and sometimes these ideas come to me as I'm doing some other task, like driving my tractor. It's a background process or subconscious process that I really cannot explain.

Usually it's related to having a problem that I want to solve. I don't sit back and dream up solutions to things that don't have a problem yet, but when faced with a problem, usually the solutions, in my field at least, come pretty quickly. I can't give you a recipe for invention. That's impossible.

Stern: Over the years, have you had any mentors and how have they helped you?

Fossum: Yes, I've had numerous mentors over the years. One mentor that I am very fond of is a fellow named Jerry Woodall, who was at IBM Research when I was teaching at Columbia. He was and still is a famous researcher and pioneer in the area of materials for light-emitting devices. He is now a professor at UC Davis, I think, and a very inventive guy. He gave me the opportunity to work with him on a consulting basis, when I was at Columbia as a professor, to come up to IBM Research and learn some new things. It was very inspiring to me to watch how he worked and brainstormed ideas. It was a very important part in my life.

Stern: Are there any inventors or inventions that you admire out of your field?

Fossum: Oh, sure. I think that one of the things you realize as you become more of an inventor is how incremental the invention process really is. That old adage that we stand on the shoulders of giants is really true. We are constantly building our technology based on the advancements and the thoughts of the people that went before us.

Stern: Does it become easier or harder knowing that, in a sense, it has all been done before?

Fossum: It's not that it's all been done before, it's that there are a lot of things that you can rely on and build upon going forward.

Stern: Has the success that you've had in this field of study changed your life at any point?

Fossum: On a personal basis, of course it has changed my life in positive ways. I was able to figure out how to put my three kids through college, which was something I was very stressed about when I was younger. I have discovered the joy of having a hobby farm and pushing rocks with my tractor, for example. But it is a little bit surreal to have an invention that was done maybe fifteen years ago or twenty years ago, that we nurtured intensely at its early stage, actually become adopted by industry.

After that, it develops a life of its own and to find it being used in so many consumer products and used so much by consumers is mind-boggling. You see how much it impacts society in terms of people sharing images and pictures on Facebook or the video news, and how fast information is shared now, for things like the Arab Spring a year ago and other political things that are captured on video. We never imagined that we would capture these things. Like the Japanese tsunami. Images from the tsunami stick in my head as well.

Stern: Is there an area that you feel that is the next usage of the technology? Or some area that hasn't been applied yet? Or an area that you're surprised that it hasn't been used?

Fossum: I'll probably not live to see it, but I think that the whole area of implantable electronics into people is really at its earliest stage. Being able to see things that are happening remotely in your brain by accessing, let's call it some sort of video stream, from afar, is going to have pretty important consequences for human evolution and society, so this kind of augmented reality is going to be quite interesting.

I think that having these low-power, lifelike, image-capture capabilities and the ability to share that imagery between people or large numbers of people is going to be very, very important. I don't know exactly how it's going to turn out, but hopefully it will turn out for the better—as opposed to some sort of Big Brother thing. It'll be very, very interesting I think.

Stern: Can you talk about the next project you are working on?

Fossum: Sure. We can consider the CCD the first-generation solid-state imaging device and the CMOS active pixel that is based on my technology as the second generation. I have been trying to explore now a third-generation imaging device that actually captures every photon that comes into the image sensor and records it in space and time very accurately.

We usually say that nature is analog, and then we have digital computers that digitize the analog world, but that's not really true with photons. When light acts like a particle, photons are very binary in nature. They are either there or they're not there, and when they hit material, they are like single points of light hitting the material. If we can capture this field of photons and then use software to create the image, it gives us a very flexible imaging capability where we can adjust the resolution, or the exposure time, or track objects much more than we can now with the second-generation or first-generation solid-state imaging technology.

Stern: Where do you think the technology will be applied? How will it be commercialized?

Fossum: I'm not sure right now. It's at the research stage in my lab at Dartmouth, but it could be useful in scientific applications first. If it pans out. It's kind of hard to say right now if it will work out or not. It could, as I said, be used in many kinds of cameras and imaging systems tomorrow, but not today.

Stern: When is tomorrow? When you say "the future," do you put a date on it?

Fossum: Well, I think I'm looking at things that have a ten-year horizon, so it's kind of hard to say. I thought that the CMOS image sensor technology would eclipse the first-generation CCD technology faster than it wound up doing. In retrospect, it's not a big difference but going forward I thought something that might take a few years wound up taking five years or ten years, so it's not always so smart to project when a technology will become compelling enough for it to overcome the incumbent technology.

Stern: Is there any technology that you've worked on that you're surprised hasn't been adapted?

Fossum: All the work that I did when I was at Columbia on high-speed CCDs, basically all that work went nowhere. I guess in retrospect I'm not surprised, but in a sense it is a bit of a melancholy feeling to look at all the efforts that we took to advance technology that actually never got used anywhere. It's always a good perspective to have when things don't quite work out.

Stern: Do you have any advice for would-be inventors out there?

Fossum: Invention is a creative activity and it is always good to come up with inventions that solve real problems, but that's not always true if you look at social computing or something like that. There is a fashion element to it as well. I think there is a difference between an invention and a product, and understanding that difference is very important. A product has to be compelling versus an invention that just has to be new. You have to be convinced that what it is you've invented is compelling before you shell out money for patents and products.

Stern: What do you think makes something compelling?

Fossum: It has to either do some existing job better or do some new job that maybe we didn't even recognize needed to be done. Social networking is an example of that and, in retrospect, I guess we know that society has always thrived on communication between people. It is something people need, but who thought we really needed it in a computing environment? So I think there are a lot of products out there, like cameras in cell phones. I didn't think we needed that way back. Wow! It's a good thing we did!

Stern: Are there any products that you really like? They don't necessarily have to be technological, but are there any things that you really covet?

Fossum: I'm not very fond of Apple as a corporation, but I really like my iPhone. It's a nice nexus of various technologies, and I find it practically amazing to have so much computing power and so much functionality that you can carry in your pocket.

Stern: What about anything nontechnological? Your tractor?

Fossum: I do think hydraulics are pretty amazing.

Stern: When you're at a dinner party and people ask you what you do, what do you say?

Fossum: That's usually a pretty good way to kill a conversation, I've found in my experience.

Stern: Why is that?

Fossum: I won't say people don't value engineers or technical people, but it's very hard to talk to other people about technical issues in a general setting. It's hard for people to relate to that, so usually they just decide that they can't understand what you're saying and turn it off. If I tell them that basically I develop technology that makes their cell phone camera work, then they get excited.

Stern: They see that as a good thing obviously.

Fossum: Because it's out there now and it's something they can relate to in consumerland. If it was not inserted into something they use every day, I think it would be a lot harder to have that conversation.

Stern: If you were to give advice considering the state of the world right now, would you say it's better to go into a corporate setting, an academic setting, a governmental setting, or an entrepreneurial setting?

Fossum: I think an entrepreneurial setting is fantastic. I think it's really the frontier in many ways in modern society, but at the same time, you need to know a lot to be fairly successful. Also, you need to have some special high-tech practical depth and you need to understand how the industry works. You can't just leap into it, generally speaking, without knowing something. On the other hand, as I tell my students at Dartmouth, it's not rocket science. You can learn a lot of the basic issues in business and innovation. They're not that complex compared to learning math methods in engineering, for example. So understanding how business works is pretty important. You have to be immersed in that for a while before you really understand that infrastructure.

Stern: What do you do for fun or distraction?

Fossum: I push rocks with my tractor. That's the biggest thing I do. I just have a hobby farm, but when you start getting into farming and living off the land and modifying the land, you really develop a deep respect for what the early American pioneers had to go through to make a new home for themselves. I guess my respect for that has grown quite a bit over the last few years.

Stern: Are you growing anything in particular?

Fossum: The joke in New Hampshire is that we grow rocks, but it's a hobby farm. We have a few animals and we grow some vegetables and a few other things, and a lot of trees. Trees are a big crop in New Hampshire.

Stern: Do you plan to retire?

Fossum: I tried to retire twice and I failed both times so far. So my coming back to an academic setting at Dartmouth is really coming out of that second retirement because I decided I still had things to contribute and I really enjoy mentoring PhD students. It's something I really get a lot of satisfaction out of and it's what I've chosen, at least for the next few years. So like I said, having retired twice, I am not anxious to retire a third time.

Ron Popeil

Inventor and Pitchman

Ronco

Ron Popeil is an inventor and TV pitchman best known for his direct-response marketing company, Ronco. Over the last 50 years, his products have pulled in more than $3 billion in retail. Born in New York City in 1935, Popeil is the son of an inventor. The first products Popeil sold came from his father's factory. Although he was never close to his father, Ron Popeil got his start in business with Samuel J. Popeil's kitchen products. In the mid-1950s, the rising medium of television caught young Popeil's eye. While he was still experiencing success with his in-person sales pitches, he saw an opportunity to extend the reach of business to the airwaves.

His infomercials made Popeil a pop icon. He popularized numerous catchphrases, including, "But wait, there's more," "Operators are standing by," "Set it and forget it," "Money back guaranteed, less shipping and handling," and "Order now. You'll really be glad you did!"

Popeil's inventions include:

- Popeil Pocket Fisherman
- Mr. Microphone
- Inside-The-Shell Egg Scrambler
- Showtime Rotisserie
- Solid Flavor Injector
- Flip-It
- GLH-9 Hair in a Can Spray

- *Smokeless Ashtray*
- *Electric Food Dehydrator*
- *Automatic Pasta Maker*
- *Ron Popeil's Olive Oil Fryer*
- *Blooming Onion and Vegetable Cutter*

In the 1990s, he joined forces with the QVC network to sell more of his inventions. In 2005, Popeil sold the company. Popeil lives in California with his wife Robin Angers and their daughters.

Brett Stern: Ron, good morning. Thanks for doing the interview.

Ron Popeil: Ah, I said I'd better get this thing over and done with. I'm swamped here. This big invention of mine just got completed after ten long years.

Stern: What is it?

Popeil: It's a deep fryer. Now, we all know there are plenty of deep fryers in the marketplace, but when I look at a category to create a product, everything I do is well preplanned. It has to meet certain marketing requirements because I'm one of the few inventors in the world who invents products on a consistent basis and only markets his own inventions. It's like a baseball player who is a great batter, but he also owns and operates a company that makes baseball bats. It's a dual role that I play here. Inventing and marketing go hand in hand.

But the invention aspect and marketing is every bit as important. If you invent a product that's not going to deliver what you expect it to deliver in the way of revenue—if that's the reason why you're doing it, you failed. Before I get involved with an invention, the marketing has to be in place. You have to understand the marketing.

You decide how many people in the world eat fried food today. Now, it's kind of funny in the North American market because we make a big deal in North America that fried food is not good for you. In the foreign marketplace—in South America, in Central America, and Asia—they all fry food. And that's all they do—fry food. And they use a variety of oils to fry the food.

I had the occasion to be with one of my closest friends in the world, Steve Wynn, and we were on his on his yacht in the Mediterranean. We were having lunch and he looked at me and said, "What's your invention?" I said, "It's a deep fryer." He looked at me and he said, "Couldn't you find an easier product to develop? The marketing on that is going to kill you."

Here's basically how I explained it to Steve:

"Before I think about inventing a product, I ask myself, 'What do I believe is really needed in the marketplace in the way of a small consumer product?'

Why a deep fryer? Deep fryers that do big foods like turkeys are extremely dangerous. But the demand for such product is high. Could I create a fryer that would do big foods and be safe? When one thinks of frying big foods, one naturally thinks that the product you create has to be big. And that's one of the challenges—keeping the product small because people have small kitchens. And, frankly, Steve, no matter how great your product is, if it's a kitchen product that's too big for the consumer's kitchen, they will not buy it. So, developing a product that is small that does big foods would be a breakthrough invention that is needed in the marketplace.

"Secondly, we've been told by everyone that fried food is bad for you. The second big challenge is to create a product that can be marketed as a healthy product. Then comes the third big challenge: to create a product that solves these two major problems for the consumer where the cost of the product would be reasonable enough for me to make a reasonable profit if I marketed it.

"My deep fryer invention is now complete. I believe I have solved all three problems. It has taken ten long years. But in TV marketing, the trick is to stay on the air by having the consumer see the infomercial run again and again. Of course to stay on the air, you have to create a reasonable profit in spite of the cost of the TV time. It is for that reason that I also have to create or find products related to my invention that gives me additional profitability to allow me to stay on TV. Always keep in mind that the more your commercial runs, the more demand you get from retail. Retail is the cash cow. But don't go to retail too soon or you might be premature and not get the results that you would get if you were allowed to get a lot of TV saturation before you go to retail."

Stern: Could we go back a little, because you're in a unique situation in that you're not the manufacturer and you're not in a particular field of production or commerce.

Popeil: Very few people are the manufacturer. If I invent a product and take it to China to find a manufacturer, am I deemed to be the manufacturer? Or is the manufacturer that makes it in China the manufacturer? Which is it to you, Brett?

Stern: I would say the person in China, but that's not really what's important here. What's important is the person getting it out there and who has the idea.

Popeil: In the way that you think, it's that an inventor invents a product here, goes to China and finds a company to make it, and he's responsible to that company. He's the one paying the bills. But if he were not, then I would not be deemed to be the manufacturer. Then more than ninety-nine percent of the people who invent product today do not manufacture their product. They don't have the facilities.

Stern: I have done exactly this, so I understand the circumstances.

Popeil: But in a court of law it might be deemed that the inventor who went to a manufacturer in China is considered the manufacturer.

One of the most difficult products that I have ever invented in my life is this particular fryer. The market is there, first of all, because of the quantity: everybody's got a kitchen and theoretically, if you look at the world as the marketplace, everybody has an opportunity and would like to use this product. But, the majority of people in this country are looking for a way to justify frying food. The marketing aspect of my fryer is that it's the only product of its kind in the world because it's designed to be an olive oil fryer. The name of my product is "Ron Popeil's Olive Oil Fryer."

When we did the infomercial on the Rotisserie—and by the way, all my previous infomercials were unscripted—the "Set it and forget it!" phrase actually came out of that unscripted rotisserie infomercial I did in the studio. I just picked up on it and kept up the repetitiveness of it, which made it a common thing across the county. When I'm in an airport, [people say] "There's the set-it-and-forget-it guy!"

This particular infomercial that I'm shooting is unscripted, except for what the audience has been told to say, and that is "Olive oil is healthier." When you hear the repetitiveness throughout the infomercial, "Olive oil is healthier," and you're someone who likes food, you're going to think, "Wait a second! That's right, olive oil *is* healthier."

Why haven't people used olive oil? Why haven't they used olive oil in North America? If you go to Spain and Italy, that's all they use.

Stern: Could you talk a little about the inventing process that you go through?

Popeil: Hold on, you're on page ten. I'm on page eight. I'll get there. Guaranteed. I'm just giving you the marketing aspect that's tied into the invention because most people do not [consider it]. When they invent, they get an idea and they proceed with the idea without looking at inventing a product and realizing the benefits from that invention. That's why ninety-nine percent of the people who are interested in inventing or reading about inventing, in creating that better, new "widget" out there need to know that you must [first think about how] you will achieve your goals, rather than spend all this money on time, lawyer's fees, and patent applications. And then you put all this money into it, and then you make a prototype and maybe you find a manufacturer to make the first piece, and then what do you do with it [when it doesn't sell]? You're stuck.

Stern: So, your process is to look at the market?

Popeil: Absolutely, absolutely. I'm just not interested in creating a product and getting a lot of wonderful patents that are worthless, except to me.

The market is gigantic for frying food. By the way, I am the world's largest collector of olive oil. I'm in the *Guinness Book of World Records*. I have a ranch where I produce my own olive oil, and so I'm a little more knowledgeable about olive oil. You don't use extra virgin, or virgin, because of the smoking point being too low. You use the regular, or pure olive oil, which is the cheapest olive oil in the supermarkets. It has a smoking point of four hundred twenty degrees.

Now, the invention itself. The big talk in this country, North America in particular, is how dangerous turkey fryers are. And yet, if you've ever eaten a fried turkey, they're delicious. The marketplace in North America is for those people who have turkey fryers, and there are two million of them. Outdoor gas fryers are the most dangerous product on the planet. Oil overflows onto your deck, and your deck catches on fire, and it's attached to your house, and it's good-bye house, good-bye deck.

So, "How do you fry a turkey in your kitchen?" was the question. The major problem in North America and around the world is that people have small kitchens. Now, if you have a small kitchen, you're not going to buy a product knowing that it does turkeys. If you do, you're not going to keep it on your counter. You're going to store it underneath your counter or you're going to put it in your garage because you're hardly going to use it. You can't sell a big kitchen appliance to someone who's got a teeny kitchen. They'll look at the product and say, "I love what it does, but it's too big for my kitchen."

Ninety percent of the time you're frying a variety of foods. If you've eaten turkeys before, you're going to eat a lot more of them now after this infomercial hits the marketplace because of the ease of the way the product works. The criteria, though, is that the Olive Oil Fryer has to be small. In your kitchen, my machine will do a fifteen-pound turkey in about forty-six minutes.

Stern: Is that the average size for Thanksgiving?

Popeil: No. The average-size turkey that is sold today is twelve and a half pounds. My fryer will do a fifteen-pound turkey. You say, "Well, what's different, when you say that it had to be small?" Well, if you look at the opening of the infomercial, it really says it all in that specific area. It basically says, "Here's a fifteen-pound turkey and here's Ron Popeil's new Olive Oil Fryer." They're right next to each other. The turkey is much bigger than the machine.

Stern: Okay. How's that possible?

Popeil: That's the invention. How do you do a fifteen-pound turkey in a machine that's smaller than the turkey?

Stern: I'm intrigued.

Popeil: Well, that's the invention. See, it also will do things that you have never done before in a fryer. Now that you have a small machine that'll do big stuff, you can do a six-and-a-half pound leg of lamb in olive oil. And then you hear the audience saying, "And olive oil is healthier." And it'll do a roast beef.

The interesting thing about frying food is that when you do a turkey, a roast beef, or a leg of lamb—and by the way, a fifteen-pound turkey uses about one gallon of pure olive oil—when the turkey is done and when the oil has cooled down, the leftover oil is the same amount of oil that you put in the machine. From a marketing standpoint, that has to really click with the American or the North American consumer. It has no effect in South America, or China, or Italy. They really don't care about that kind of stuff over there.

Stern: A majority of your sales are in North America?

Popeil: Over ninety percent have always been in North America, but that's from a choice of not wanting to market product in Europe and Asia. Instead of making billions of dollars, I made hundreds of millions.

Stern: Do you think that growing up you were an inventor-type of person?

Popeil: I was. I started out as a marketer and salesman. I started marketing and selling my father's products. But I was a customer of his. I was never an employee of my father, who invented a variety of products. I would buy them from him. I would pay the same price that other people had paid, except I did have one edge over everyone else. He gave me credit. Ninety-nine percent of his customers were basically pitchmen at county fairs and state fairs, home shows, and things of that nature. Woolworth stores. I did pretty much the same thing. And then I stopped buying my father's product, and I started inventing my own products and marketing my own inventions.

Stern: Can you discuss the development process for this current project?

Popeil: Normally, for a patent application on a product like this would be less than a quarter of an inch. This one is over two inches. So, you get an idea that there is a lot of intellectual property tied into the product. But that's something else. Those are little things that make the product work, and that you need to do to accomplish the job that you're intending to do.

When I went to Underwriters Laboratories, I met with three engineers. I showed them how the machine worked, and the head engineer came over to me and he whispered in my ear, "Ron, that's the greatest machine that I have seen in many years." He said his engineers both want to buy one.

If Underwriters Laboratories or any electrical agency has regulations, and you meet those regulations, they have to give you their approval. Well, I met all the old regulations, but now they're changing their regs. So, it's another problem you have to deal with when you create an electrical product. If you have any problem whatsoever, the judge will throw the book at you because you didn't get an electrical approval. So, it's mandatory before you sell the first piece.

Stern: Wouldn't you say having these hurdles thrown at a project is typical for an inventor?

Popeil: No. This is above and beyond. Unfortunately, of all the products that I have invented that have needed Underwriters Laboratories or electrical agency approval—a Pasta Maker, the Food Dehydrator, the Showtime Rotisserie—all those were a slam dunk.

Stern: I meant just the hurdle of something put in front of you, as an inventor?

Popeil: Yes. You're absolutely correct. The same problem, whether it's easy or difficult, you still have to deal with manufacturing, approvals, safety issues, design, marketing.

Stern: What do you think are the skills sets required to be a successful inventor?

Popeil: Number one is common sense. You have to have an understanding of what's going on in the marketplace. Many inventors are recluses. They don't know what's going on out there. And yet, sometimes they're able to create. Now, guys like me who are what I call "common sense inventors" are way behind when it comes to the kind of inventions that are taking place in the technical world today. I may be experienced in the real marketplace, but the real money to be made today in inventing is in the technical area, of which I know very little about.

Stern: When you're doing the actual inventing, do you collaborate with anybody?

Popeil: Yes, I collaborate with an individual who is an expert in engineering. He's a minority partner of mine. He is not a patent lawyer but thinks like one. He basically writes the patents in conjunction with the patent lawyers. That's another area of inventing that your readers should know something about. Your readers think that the patent is a patent is a patent, and it's only protection, and it's not that way. You can take it to ten different patent lawyers and each lawyer will write it up a different way. Some will write it so it's circumventable. Some guys know how to write patents, and it's the choice of words that make or break your patent.

Stern: You do your marketing first, and then you get to the point where…

Popeil: That's only done in your mind, though. You have nothing to sell. What problems do you have to solve? Can you really fry in olive oil? Well, the first thing you have to do is to test the product. Now, you don't have the tooling to make the product. If you did, the product would be totally designed already. It's a function of prototypes. One of the things you see in the infomercial this year is all the prototypes that I have worked with to get to the size perfect. I have prototypes that are four times the size of the machine.

The first thing you're going to have to do no matter what the invention, is have someone make one for you—or make two, or three, or twenty—so you can test it to see how you can improve it and make the necessary changes before you go into the manufacturing aspect, where there may be a lot of tooling necessary. But that's after all the research and development has been done.

Stern: Do you start with a sketch on a napkin? How do you even get to that prototype stage?

Popeil: You're asking how prototypes are made? They're made in a variety of different ways. Is the person inventing the product capable enough to draw something on a napkin? Or does he have an idea and takes it to creative people who actually produce prototypes for people based upon their ideas? It starts in the brain. It ends up on paper, either from you or from whomever you take it to. Whether it be the kind of material that you're going to use on the product or a different kind of material, but contains the same shape and acts in a similar fashion to the kind of material you're going to use going forward.

Stern: When is the Olive Oil Fryer coming out?

Popeil: That's another issue. I sold my business ten years ago and the company who bought it from me had the right to purchase this particular invention at the beginning stage for $10 million on a given day. And if they didn't purchase it for that amount, they lost all the rights to it. So, my situation was I ended up selling my business with all the products and the intellectual property for just under $60 million. The bottom line is they missed out on the new product.

When I invent a product like this, it's not just one product. It's a combination of products because the marketing has to come into play. I need the profitability to stay on TV, so when you buy my fryer on television [you'll learn it does more than fry meat]. If you put water in my machine, it converts into a big food steamer, so you can do twelve pounds of frozen seafood in around twenty minutes.

Stern: Wow!

Popeil: That's part of the marketing at the end of the infomercial: If you order now, I'm going to give you this seven-piece food steamer attachment that's collapsed now, but when you stack them up, you can do your king crab legs. When I say "a food steamer," it's quite unique in design, as well. It's in a collapsible form to begin with, so it doesn't take up a lot of space when stored.

Stern: Do you have a timeline for when you plan to hit the market with this?

Popeil: Well, in the first place, the infomercial has to be done. And that was shot July 12th [2012]. I also have my own cameras, my own lights. I shoot a lot of the stuff, and I do my own food styling. I do it all myself. I do everything here. I write the booklets for the print material. All the instructional material, I create it myself. I write it myself. I do all the food shots. I do the food styling, I do the cooking, and I do most of the shots myself. By the way, I'm using my iPhone a lot. The camera is pretty good on those.

I have to find someone to buy the project from me, the whole thing. I'm not going to go back into business—have two hundred employees like I used to have, customer service, and insurance, and everything that goes along with operating a business. I've already had one group of people look at it. They offered me about a $16 million package and I walked away from it because it wasn't enough money. They had agreed to all my terms, but the money wasn't sufficient.

Stern: Why do you think that the company that you sold your organization to didn't want to participate?

Popeil: They couldn't raise the money. They could raise the money if I went out with them to help raise it. And I said to them, "Wait a second. You want me to go out and raise money for you to give to me? Oh, no, no, no. It doesn't work that way."

QVC came out to my house to look at the prototype. The guy said to me as soon as he saw it work, "I want to buy fifty thousand for a one-day sale run, and I'll buy another fifty thousand for the balance of the year." That was his initial statement to me. They call now every couple of weeks asking, "When will you let us know?" I will not give them a date because of the Underwriters Laboratories thing. I cannot, because of the way I kind of do business where integrity is paramount.

Stern: How do you evaluate a product being successful?

Popeil: I only do it one way and you know the way. How much money did it generate? I'm sure if you read any biography or have seen stuff over the years, you know that I've been successful since the first day I went into business. Success to me is "I'm only as good as my last product." This is my last product. The question is how much money are you going to make? Are you going to make hundreds of millions? Are you going to make just $20 million? I do not get involved initially with an invention if I can't make tens of millions, if not hundreds of millions of dollars. That is the sole thing. I am not interested in having the best-looking, best designed piece of product that sits on a shelf and collects dust.

Stern: In looking back on the list of inventions that you've put out there, certainly some have been more successful than others. I'd like to ask what has been your biggest failure. Why do you think it failed and what did you learn from that failure?

Popeil: If you think making a half million dollars is a failure, and I don't know who would, if you're a corporation that does hundreds of millions and you make a half-million dollars, that could be deemed a failure. But, if you're an individual growing up, and you haven't gone to college, and you've never worked, to me, a half-million bucks is still a half-million bucks.

I can't remember the last time that I made a half-million dollars on a project, though. In the earlier days, that certainly was the case, but was there a product that I lost money on? No. I've never had that kind of thing. The question is, are you going to make a couple hundred million? Or are you going to make $20 million? And those are the kinds of things I deal with. Most of my inventions' sales run anywhere around a couple hundred million. The Food Dehydrator was over a half-billion. The rotisserie was $1.3 billion, $1.4 billion, and so the Rotisserie is the biggest project that I have ever had. My knives did big business and still are.

Stern: Are there any inventions out there in the marketplace that you wish you did?

Popeil: Velcro. I had a home in Aspen on the mountain and we had those little things growing out of the ground that would always stick to your jeans and socks. It was there in front of me all the time—and I missed it!

Would I have liked to done The Clapper ["clap-on/clap-off" switch]? Yes, I would have. I get credit for doing it, although it wasn't my product. I thought that was a very clever item.

The George Foreman Grill, I had something to do with. Most people don't know that. It was another mega-bucks product. The manufacturer came to me with it and it was more of a marketing thing. The product had already been invented. Even when they hired George Foreman and gave him a hundred thirty some-odd million dollars for his name and face. And he knew nothing about the product at all. Be that as it may.

Stern: Going back to Velcro, you mentioned the inspiration for that idea. Where do you find inspiration?

Popeil: To me, my inspiration is somewhat mundane. It's just driven by whether I can select a category to move my effort, time, and money into something that will produce the result I'm looking for, which is strictly tied into the monetary aspect of how much money I can make with my invention. If you make no money with your invention, basically you've really failed in the real world.

Stern: Earlier you said one of the reasons you're doing this interview is so inventors could learn some knowledge.

Popeil: Knowledge about the marketing aspect. Look, the American amateur inventor sees a lot of people on television promoting, "Send us your invention. We'll help you get it off the ground". Well, you know that ninety-nine percent of the people who send these companies money never achieve the goals that they thought they were going to achieve. [Those services] don't give a damn that your product is not marketable. They'll just take your money.

Stern: The individual inventor shouldn't use these outside design services or inventing services?

Popeil: You've got to work within the talents that you have. If your ideas are good, and the marketing aspect is good, but you don't have the ability to make prototypes, then you've got to do some research to find out where you can. Manufacturers everywhere in the world, especially in Asia, have access to making prototypes. If you run up against a problem in North America such that you can't get someone to make you a prototype the way you want, then get off your butt and go to China.

There are different shows/conventions that your product area might fit into. You go to one of those functions to find out who's making what. Tie yourself in with a manufacturing company that has engineers. Basically, most all of them know how to do prototypes. You get the benefit of the cheap labor. We make a lot of our own prototypes here, ourselves, and when we can't, we'll go outside the box. We'll get the manufacturer to assist us.

Stern: You sold your business and "retired." What was the incentive to do one last project?

Popeil: The product had already begun. The company had the right to buy it, but when the day came for them to fund it, they didn't have the money, and then they lost all those rights. So, what do I do with the project? It's the early stage, but it has already gotten off the ground. Am I just going to walk away because this company that bought my company couldn't do it? I have enough resources on my own to complete the project. It's "the game" of whether you can achieve the goals that you want. I don't have to work anymore.

Stern: Could I ask what you do for fun or a distraction?

Popeil: Well, right now, there is no fun. Infomercial day was one of the most difficult days of my working career because I'm the only one that knows what's going on. There were two hundred in the audience, a studio, eight cameramen, electrical, and food. We have to look good. I've got turkeys and I've got olive oil. It's one hell of a day. So, my focus is in the attempt to sell this project to someone.

Stern: But, wait, you've had this time off over the years. You have to do something for fun?

Popeil: There is no time off when you're working. I don't have time off working on a project. Now, if the infomercial is not around the corner, you're trying to put everything together: the booklets, which I have to do, and the packaging. Everything is done by me. Anyway, the bottom line is, I want this whole thing done some time right after the infomercial. It takes a month to edit and then sell the project to someone. But right now, to answer your question, when I have a couple of days and I can't do the kind of work I'm attempting to do, fishing is my hobby. I have a boat called *Popeil's Pocket Fisherman* and I go fishing. That's the hobby.

Stern: Do you use the Pocket Fisherman for fishing?

Popeil: The answer's, "Yes, I do." I kept my boat up in Alaska for ten years. It's a custom-made boat and the first time I used the Pocket Fisherman, a king salmon got on and he ripped the goddamn thing to pieces and out my hands. And to this day, there's a fish, and a line, and a Pocket Fisherman swimming around Alaska. No, I do not use the Pocket Fisherman up in Alaska. I do use it here for rockfish near the Channel Islands, but not up in Alaska. Forget about it. Those fish are just too big for my rod.

Stern: Any final advice that you have for a would-be inventor out there?

Popeil: Well, as I said, the first is primary: before you spend one dime or one minute on your project, get an idea of the aspects of how this product is going to be marketed and the real benefits of it. Don't come out with some product that you get a patent on, but there's twenty other items that do the same thing and they have their own patents. In that situation, the market has really been shrunk down to a minimal size. You've got to really understand what is really unique. If you're going to be the only guy in town, then you really have something.

For my Turkey Fryer, no one has got a product like this out there, so it makes a great gift for Christmas. It makes a great gift for anniversaries and weddings, holidays, and that nature. You, yourself, don't have to be the user of the product. You make a product because other people want to buy it as a gift for somebody. And so, that's the benefit of creating something unique. But if you create something and you get a patent on it, but the patent you got is unique to the product, but the product itself is not unique, it's a marketing choice.

I do hope that your audience and your amateur inventor out there really, before he puts the effort, money, and time into it—*effort, money, and time*—gets a good handle on the marketing.

The success of the marketing is due in part to the upsells. Maybe you invent one product that is not enough to be successful on the monetary end. You have to create a series of products related to that core product. I make the best Blossoming Onion Cutter in the world that's an upsell with [the fryer]. It also makes french fries, and slices and wedges, but it's an ancillary product. Once you've ordered the fryer, and we have your credit card information, we're then able to offer you the sale of this other package. Sixty percent of the people on the rotisserie took that package, and if they take the package, then you have a second upsell solely for those people who took the first upsell. So, there's a knowledge of marketing that helps you achieve the goal for the kind of profitability you're looking for.

By the way, my audience of two hundred people, each got a machine, a turkey, a gallon of olive oil, and my new Professional Turkey Carving Knife. They used the product before they came to the shoot on July 12th. Everybody in my audience used the product. I believe in honesty in marketing, so the testimonials you've seen in the past are real. It adds up to the final goal of achieving what you think is going to make you successful in that arena.

Stern: Ron, it's a great story. I think it will be incredibly inspirational and motivational to people who are going to read this and follow some of your reasoning to go forward.

Popeil: You know, it's a strange thing, but in the housewares business, in particular, all these big companies, they don't create anything. They just make another rice cooker. They may make it bigger, make it smaller, make it another color, but nobody wants to invent.

Stern: Most organizations are certainly risk-averse.

Popeil: Exactly. Exactly. You hit it right on the head. They don't want to put their money where their mouth is.

Robert Dennard

IBM Fellow

*Born in Terrell, Texas, in 1932, **Robert Dennard** earned undergraduate and graduate degrees from Southern Methodist University (SMU). In 1958, he received a PhD from the Carnegie Institute of Technology in Pennsylvania. The young engineer and scientist soon joined IBM's Research Division and eventually began working on integrated circuitry.*

In 1966, Dennard conceived of a revolutionary approach to computer memory— dynamic random access memory (DRAM)—and received a patent for his invention in 1968. In 1974, Dennard and fellow IBM team members published Design of Ion-Implanted MOSFETs with Very Small Physical Dimensions, *which set forth guidelines for long-term advancements in RAM. The document is regarded as a seminal reference in the field of computer hardware. Dennard's breakthrough technology and innovative concepts enabled the evolution of modern computing, as well as the proliferation of today's sophisticated electronics products.*

Throughout his half-century career at IBM, Dennard was issued over 60 patents and won numerous honors and awards. Appointed an IBM Fellow in 1979, he received the US National Medal of Technology from President Reagan in 1988. He was a US National Inventors Hall of Fame inductee in 1997, and received the IEEE Medal of Honor in 2009. Dennard and his wife, Jane, enjoy choral singing and Scottish dancing.

Brett Stern: What is your background? Where were you born and what were your education and field of study?

Robert Dennard: I was born in Texas and, considering my age, that was a totally different world then. This was in the middle of the Great Depression of the early thirties. I grew up in a rural environment until the forties, when World War II changed that situation and my family moved to a town just outside Dallas called Irving. I lived there and completed my high school career. I went on to

SMU, which was nearby, to study electrical engineering. I continued my education at Carnegie Tech, which is now Carnegie Mellon University in Pittsburgh. I graduated from there in 1958, ready for a career.

Stern: Would you consider yourself a good student?

Dennard: I was an excellent student. However, it was a lot different in those days. I just studied the things that were provided in the educational system, pretty much. I had some opportunities, starting in a one-room schoolhouse, to have an open kind of classroom environment with three classes in the same room, which was interesting. That rural schooling went on through the fifth grade. So, I was always able to hear what was going on at the higher levels at the same time, and I advanced pretty rapidly, skipping a grade and getting into college at age seventeen.

Stern: How did you pick electrical engineering?

Dennard: I picked electrical engineering on the basis of a discussion I had with a guidance counselor at my high school. She noted that I had a good mathematics aptitude and was interested in science, and the fact that engineering was a very good field to go into. It had a lot of demand at that time and that sounded interesting. So, I bought a slide rule and went off to college.

Stern: When you were in college, did you invent anything there?

Dennard: Not really. I did work on a co-op job at SMU, which was good and it allowed me to earn some money. I was in school half-time and working half-time for Dallas Power and Light Company. I did come up with an idea for improving the regulation system for transmission lines.

Stern: Any thoughts on how a cooperative education prepared you for your career?

Dennard: Mostly it provided income. It gave me a year to grow up and a chance to interact with people. Getting along with people of all varieties is a pretty important thing.

Stern: When you got out of school, you had a degree in electrical engineering?

Dennard: Yes, I had gotten all the degrees I could get in the subject, culminating in a PhD from Carnegie Tech.

Stern: Was IBM your first position when you got out of school?

Dennard: Yes. I joined IBM, and I've been here ever since.

Stern: When you started, what was the job description? Did you think you were going to be an inventor, or was it going to be more dealing with day-to-day things?

Dennard: I was definitely hired into the Research Division, and it was pretty clear that we were searching around to find out how to make computers better.

It was very early in the computer era and IBM was hiring a lot of people because they figured this was going to advance rapidly. We were looking for the ways to make that happen.

Stern: What year was this?

Dennard: 1958.

Stern: You are known as the inventor of DRAM [*dynamic random access memory,* pronounced "DEE-ram"]. Could you provide a definition in terms of the technology?

Dennard: The technical part is a very simple explanation. I have to distinguish what was there before. Previously, from the early forms of semiconductor memory up to the time of this invention, the RAM that was being used was magnetic-core memory. This technology was a bunch of little, tiny magnetic toroidal "beads"—or I like to think of them as "Cheerio" shapes—that were put together with wires that were strung through them to make them a way of storing maybe thousands of bytes in a planar array. It was quite a large plane, maybe a foot wide.

My invention really moved into a totally different field—a semiconductor field based on transistors, which were beginning to replace the logical functions of computing. The first memory built in that technology used transistors in a bistable flip-flop circuit, which had one inverter feeding another inverter with a feedback path. It took six transistors to store each bit of data—whether they be bipolar transistors, which was the mainstay of computer technology in those days, or field-effect transistors, which was the alternative technology that I was exploring, along with a fairly large group of IBM researchers.

One day, I was stimulated to look for a much simpler approach by a talk I heard from another research group, which was trying to extend the magnetic-core technology. I became aware that they had a simple element—this little piece of magnetic material—that actually did the storage. So I started looking for something in our field-effect transistor technology that was analogous to that or had that same kind of property.

We knew how to build capacitors in our technology, so I came up with the idea that a capacitor could be charged to two different voltage levels to represent digital information. After a couple of months of exploring ideas in my spare time, the DRAM memory cell that I came up with is just a single capacitor and a single transistor to store each bit of data. The transistor is used to steer the desired voltage level into this capacitor from a data line and to read it out of the capacitor back to the same data line. I found a way to build a large array of memory cells composed like that, with address decoders to turn on one row of transistors at a time and to connect selected data lines to the input or output terminals.

Stern: How does the public perceive the technology?

Dennard: They know that their computer has RAM in it, and that it's measured in gigabytes these days.

Stern: From the time when you had this inspiration, what was the approximate timeline of the process that you went through—from prototyping and experimenting to the point of where you got something that was actually working?

Dennard: It was a little bit different than the usual process. This was really early in the field-effect transistor technology. It was just being developed, and so the primary thing I did at that time was to show that a very simple circuit in the laboratory—just a single capacitor and a transistor—could do this, and to measure the waveforms that were there. The invention was completed in late 1966. I filed for the patent in 1967. It was granted in 1968. A very fast process. There was no field that had to be searched. This was a totally new field.

Stern: So there was no prior art?

Dennard: No. I have a book on DRAM inventions—*Inventions in the Field of DRAM*—and mine is the first one in the book. It went on from there. It was really early in developing this technology. Our first products were using those six-transistor memory cells that were already being developed.

Stern: Did IBM do the actual manufacturing, or was this a joint venture with another company?

Dennard: IBM did its own development and manufacturing of semiconductor memory technology.

Stern: Did you do your work by yourself, or was this a collaboration with other researchers in the lab?

Dennard: I was working in a program with a small group that was exploring the circuit opportunities of this field-effect transistor technology, but I was the one in this group who was particularly looking at the memory issues. I was the only one working on this novel idea that I came up with.

Stern: Could you explain your ideation process? How did you define problem? What steps did you go through to analyze them and come up with this solution?

Dennard: The process starts by asking questions. You look at something and you say, "Why is it done like this? Is it something that could be better? How does it really work? And why not do it some other way?" It's just a lot of questions like that.

Stern: When you come up with these questions, do you give yourself many possible solutions? Or do you just pick one and then go down that road?

Dennard: Well, hopefully, something pops up in your mind. It could happen when you're asleep at night or early in the morning, after you're rested a bit and your mind is more aware that you're going through some kind of synthesis. All of these loose ends that you have in your brain, you start examining, and things become very, very clear.

Stern: It sounds like you get these solutions when you're in somewhat of a relaxed state of mind?

Dennard: Absolutely. It takes a slow process.

Stern: Do you visualize these solutions in your head as a finished idea, or as an "exploded view" or as a drawing?

Dennard: It's abstract. It doesn't seem to me to appear as a picture. It's more abstract thoughts. If it is a circuit idea, sometimes I've gotten up in the middle of the night to write down something about it, or maybe to draw its topology.

Stern: Do you use any specific type of tools when you're thinking of ideas? Do you sketch, prototype, or model them in any way?

Dennard: If I'm awake and at my desk, I have a large bunch of loose papers on which I start drawing little pictures. I've got a blackboard full of drawings with many, many subjects on it. I don't erase it. I kind of leave those things up there. So, if I need to erase something, I just erase enough to add something new. I try to figure it out.

Stern: You just said that you look at other subjects. What other subjects outside the electrical engineering field do you find helpful to stimulate ideas or influence your thinking?

Dennard: Well, the field I'm in is very large. It includes physics, chemistry, integrated circuit processing, and device structures as well as circuit ideas. I've used really fundamental engineering concepts to analyze what I'm doing. I have techniques to solve circuits or answer device questions on the back of an envelope. It's always good to be working with some associates who have the capability to delve into a computer and model it with great precision and great accuracy as long as it's consistent with the assumptions of what has been put into the computer. I think the ability to think through the problem and roughly get quantitative solutions is pretty important to an innovative person.

Stern: In general, do you like working alone to come up with some possibilities, or working in a collaborative effort from the beginning?

Dennard: I like working collaboratively from the beginning. I'm currently enjoying my situation as an IBM Fellow. I generally have one or two people that I'm collaborating with on an idea, and pushing it along and trying to get it adopted by someone who can use it.

Stern: How would you say your invention has influenced society?

Dennard: It's a totally different world today—not just because of my invention. My invention is just one step in the evolution of the transistor into microelectronics and integrated circuits through many, many important steps and contributions from thousands of people. It's not just how DRAM has changed society, as much as how miniaturization of semiconductor integrated circuits has changed society. The scaling principles, which I worked on with several associates, also helped drive that miniaturization progress.

Stern: Can we talk about the scaling issues?

Dennard: Let me describe how we got there, which is related to the DRAM. After we had finished our research on the field-effect transistor technology and transferred it, we went through a period here where we thought that we'd done our job and semiconductors were conquered. That was it and that would be the technology forever, right?

We went off and did some new stuff in communications that was the next big thing. As things started evolving with the manufacturing of semiconductor memory, people became aware that a lot of memory was going to be needed. Fantastic quantities. How were we going to supply all those quantities of memory?

We were building these things in those days with two-inch-diameter silicon wafers—that's why they called them "wafers." We would have to open hundreds of plants all over the world to manufacture those in the quantity that we would need to ship supplies to an exploding market for computers.

Some smart guys here at our research lab and some of the management came up with the really good idea of starting a new project to see how we could drastically reduce the cost of semiconductor memory. They perceived that the increase in productivity required to reduce the cost would enable them to be produced in great, great volume. We had a goal set up for our engineering group to come up with a solution on how to get memory down to one millicent per bit, which was quite a large jump from where we were in those days.

We had some ideas on how to do that because we had some good projects in electron-beam lithography. We had systems that could write patterns for integrated circuits at much smaller dimensions, and we also had optics people who thought we were far from the limits of optical lithography. The idea was to make things much, much smaller.

Also, at that time, the one-transistor DRAM cell I had invented had not yet come into manufacturing. This was early in 1970, and these ideas presented an opportunity to have a much simpler memory cell with smaller dimensions that could have many more bytes on a chip.

To do all this forced us, basically, to invent the principles of scaling. How are we going to make them smaller and make them work? When we really examined those questions, we saw other questions. After a few days of thinking about it, we found a simple solution: make all the integrated circuit and transistor dimensions smaller, lower the applied voltages, and increase the doping concentration in the transistors. This solution reduced the depletion layers between the P-N junctions at the same time as everything else was being reduced.

Stern: It sounds like the scaling process came about because you made the memory. It's almost a chicken-and-egg thing: if you didn't have the memory, you wouldn't be doing the scaling?

Dennard: That's how it came into our thoughts. And, as it turned out, within a few years, the first one-transistor DRAM chips started being produced and really drove the future scaling process. It was a high-volume, very reliable product. Moore's Law, as it came to be known, is the way this miniaturization proceeded with time, and it was very much DRAM that was driving that process for many years.

Stern: Do you think it's just going to keep getting smaller and smaller and faster and faster? Are there no limits?

Dennard: No. We've been working on low-hanging fruit for about thirty-five years and all the low-hanging fruit is gone.

Stern: Did you think it was low-hanging fruit when you picked it thirty-five years ago?

Dennard: Not at all.

Stern: What do you think the future or the next generation of your technology will be or look like?

Dennard: We want to continue the scaling trends as much as possible, but now we have gotten to the point where we have to do something quite a bit different. The industry is pretty much now going into transistors that are made in very thin slices of silicon. This can be done with a silicon-on-insulator process, which has been around for about twenty years. But now we would make this silicon layer above the insulator on the silicon substrate ultra-thin to improve what we call the "short-channel behavior" of the devices. This effect enhances performance in this device by improving its electrostatics, so that when we make it smaller, it will still turn on-and-off properly.

Stern: So, there's still development going on?

Dennard: The new FinFET technology—in which the ultra-thin layer is produced vertically by etching silicon into very small, thin slivers called "fins"—is just coming into manufacturing.

Stern: Could you talk about any failures you might have had, and what you learned from these failures?

Dennard: Wow. I've been pretty lucky. What might be considered failures are actually other great ideas. Maybe they just don't have the impact that DRAM and scaling had, so they haven't spread. What I've learned is that it's difficult to get incremental improvements accepted. People didn't want to take on the change involved, I guess, because the old ideas were still working well enough.

Stern: Could you talk about the intellectual property side? Any thoughts on the way you secured patents and the value of them? Any advice for people trying to get patents?

Dennard: I don't really have a lot to say on those questions because I'm not involved on the business side. Obviously, the DRAM patent has been a very important part of my company's intellectual property. I do have about sixty patents, and I'm still contributing and working on more.

Stern: Can you talk about the project you're currently working on?

Dennard: I'm very involved in looking at those new device structures that we talked about, and how to build very energy-efficient circuits using them.

Stern: Do you think there are any particular skill sets or personality traits that someone needs to be a good inventor?

Dennard: I've thought about that, and I've thought about creativity and how and why is it that some people are more creative than others. In my experience, creativity is not a widespread phenomenon. In this research lab, amongst all the people that I've worked with or know about, some are much more creative than others. One trait that distinguishes the creative people is their attitude. When they get a good idea, they believe that it could really influence what's going on in the world and make a difference.

Stern: Do you think creativity is something you're born with, or is something that could be developed?

Dennard: Probably attitude alone is not sufficient. You might need something else there. I think that people who are creative have the capability to access a lot of things in various areas of their brain and somehow put them together. I sometimes speculate that, thanks to being brought up in a rural area, I got plenty of time to develop my brain at a very slow pace with no distractions. I think people are being brought up these days under exactly the opposite conditions, with so many things being thrown at them. It might be that they'll end up being the ones who are much more creative than I am, but I think my ability lies in being able to think very slowly and just contemplate.

Let me just mention that at a reception one evening at the Inventors Hall of Fame, where I'm privileged to be an inductee, I gathered with a bunch of my fellow inventors, and we discovered that we had all grown up in very quiet rural settings with very small schools, predominantly in one-room schoolhouses.

Stern: In reference to your fellow inventors, are there any inventions or inventors that you admire?

Dennard: Gee, I have to be careful. After all, they're going to read this book. I admire them all, but the one I've admired most recently was just last night, as I was riding along in my wife's car. Her GPS was turned on, and I was looking at the map of the car moving along. And I said, "You know, this thing's being mass-produced for very few dollars, and it's everywhere. There's one guy, Bradford Parkinson, who really developed all of that." It's an amazing thing.

Stern: Do you have any advice for would-be inventors?

Dennard: Inventing is actually very hard work. Many times, I have gone to bed at night glowing about a great new idea, only to wake up the next morning realizing why it won't work. But the satisfaction of those few ideas that really succeed is worth the work.

Stern: Do you feel that you have to be the "cheerleader" of a technology that you invent? Do you feel that it's your responsibility to continuously push the project forward?

Dennard: I do feel a deep personal responsibility for both DRAM and scaling. I made many trips to our manufacturing plant in Burlington, Vermont, as IBM was developing the first product chips using the single-transistor DRAM cell. As we moved from 64-kilobit chips to 256-kilobit chips to one-megabit chips and so on, I was following all of that and thinking ahead, where things were going and what was going to happen.

It was the same thing with scaling. I've looked ahead and anticipated what the problems would be as we scaled voltages to lower and lower levels, and I helped develop solutions to those problems. The scaling rules have actually been generalized and modified over the years, as I've learned new things and learned to express the scaling laws in different ways.

Stern: Can I ask you what you do for fun or a distraction?

Dennard: I do choral singing, and Scottish country dancing, and other forms of dancing.

Stern: You've been at IBM now for forty-odd years?

Dennard: Fifty-four years, I believe, on June 16.

Stern: Do you plan to retire?

Dennard: I don't plan to retire, no. I haven't time to retire.

Stern: Are you going into work every day?

Dennard: Yes, indeed.

Stern: Any parting advice you want to give to inventors?

Dennard: Yes. The main thing to remember is that anyone can do it. Curiosity and a can-do attitude are very important. And it's extremely helpful to be in the right place at the right time [laughs]. There are still a lot of opportunities left and a lot of key problems facing us, where someone can make a big difference.

Gary Michelson

Orthopedic Surgeon

Dr. Gary Michelson was born in Philadelphia in 1949. He attended Temple University and Hahnemann Medical College. Dr. Michelson completed his residency in orthopedic surgery at Hahnemann Medical Hospital. He completed a fellowship in spinal surgery at St. Luke's Medical Center in association with the Texas Medical Center and Baylor University.

Dr. Michelson is a board-certified orthopedic surgeon and a diplomat of the Academy of Orthopedic Surgeons. His inventions cover more than 250 United States patents—and more than 950 issued patents or pending applications throughout the world—related to the treatment of spinal disorders and surgery, including instruments, methods, and devices relating to advances in spinal surgery and the general field of orthopedic surgery. In 2005, he sold ownership of many of his spine-related patents to Medtronic for $1.35 billion, propelling him onto the Forbes 400 list.

He is the founder and sole benefactor of the Medical Research Foundation Trust, the Michelson Medical Research Foundation, the Found Animals Foundation, and the Twenty Million Minds Foundation. Through his foundations, Dr. Michelson has donated over $100 million for cutting-edge medical research. The Twenty Million Minds Foundation was created to provide for the use of a comprehensive library of higher education textbooks in an open architecture platform to replace the costly textbooks-for-purchase system currently in use. The $25 million Michelson Prize is being offered to address the issue of pet overpopulation. The prize is for access to a single-dose, safe, and effective sterilant for cats and dogs. Michelson has provided an additional $50 million to fund medical research in support of the prize.

Dr. Michelson funds entrepreneurial competitions at a number of business schools, including The Wharton School of the University of Pennsylvania, Stanford University, and the Massachusetts Institute of Technology (MIT).

Dr. Michelson is a board member of the Educational Foundation of the Intellectual Property Owners Association. In 2011, he was inducted into the National Inventors Hall of Fame. He received the 2006 Paralyzed Veterans of America Award for the Outstanding Medical Research in the field of spinal disorders.

Brett Stern: Please tell me about your background. Where you were born? And what were your education and field of study?

Gary Michelson: I was born in Philadelphia, Pennsylvania. I went to Temple University for college, then Drexel University, which was then named Hahnemann University, for medical school. I did a general residency in ortho-pedic surgery, and then I did a fellowship in spinal surgery in a special program in Houston, Texas, that was a combined program between Baylor and the University of Texas. Then I moved to California and practiced orthopedic surgery and, predominantly, spinal surgery.

Stern: Growing up, would you say you were an inventive type of child?

Michelson: In those days, you could, as a kid, walk around and find rather intriguing things put out for the trash man, such as phonographs. I would take those back home, take them apart, and usually it was pretty self-evident why they didn't work. For the most part in those days, it wasn't too difficult to fix things. You could improvise parts, or find a nut and a bolt and put something back together.

I guess along the same lines, I remember being sick one time and staying home. Of course, you were not allowed to watch television, because that would be rewarding, but I did watch television. I was watching television and the televi-sion stopped working. I figured, "That's not going to work out well. I'll be in big trouble." So I took the back off the television set and, interestingly enough, that's even before they had automatic electrical disconnects. If you took a back off a television set ten years later, the plug unplugged. But back in those days, that didn't happen, so you could actually take the back off and have the entire televi-sion set still electrified. You'd look in there and see all these glowing tubes. Even as a kid, you'd see one that wasn't glowing. You really didn't have to be a genius.

So you pulled the tube out, and went down to the five-and-ten. They had a kiosk with a tester on the top and new tubes inside. You test your old tube and if it isn't working, you buy a new tube, put it in your TV set, and your TV is working again. I was in the middle of doing that when my brother got home from school at lunchtime. He was about a year-and-a-half older than I was. He says, "What are you doing? You can't take apart the TV set." I tried to explain to him what I was doing. But he said, "You're in huge trouble." Then I put it back together and, of course, it did work, because all that was wrong was the tube.

So the other thing is—and it's probably not exactly what you asked, but you're going to get to it anyway—why do certain people think in certain ways that seem to be different than the ways other people think? All graduating high school students get to take something called the merit exams. For a very long time, they've been tracking exam-takers' birth order—whether you're an only child, the dates and the relationships between your siblings—and they have published this information.

It turns out that most merit scholars are only children, and if they're not only children, then they're firstborns. It gets harder and harder, apparently, for a child to succeed academically if it has a number of other older children. That same information is reflected if you look at the people who were United States astronauts. I think out of the first thirty, they were all either firstborn or only children. Now can it be that the firstborn or the only child somehow is geneti-cally superior or more intelligent? Well, you know that can't be right.

Stern: Is it environmental?

Michelson: Well, sure. There's something else going on, and it's interesting that nobody has really tried to delineate what that is, because it's powerful and the answer is pretty easy. If you look at my own family, all the people there are reasonably intelligent. So when my brother—my older brother, the firstborn—was three or so, my mother's sitting there. She's a young woman. She's bored out of her mind. She can't go anywhere because she's got the kid. So she teaches him to read, and print, and write, and do math, so by the time he starts kindergarten at five, he's got the education of a second-grader.

His life experience is that every teacher who gets her hands on him thinks it's reflecting their greatness as a teacher, and they love him and they think he's brilliant.

So I come along. Now my mother didn't do any of that for me and, in fact, and this is just a personal anecdote, my father actually used to call me stupid. "Come here, stupid. What are you doing, stupid?" I have a very clear memory of that. I actually thought my name might be Stupid. That was his way of making himself feel smarter. But out of that dynamic, I got to school and these teachers actually would say to me, "What happened to you? You're an idiot." They would actually say it right to my face: "What happened to you? You're an idiot."

Stern: How did you get over that?

Michelson: Well, I never got over it while I was in school.

Stern: Was it a motivation for you?

Michelson: I can tell you, I only graduated high school because the teachers could not stand the thought of having to put up with me for one more year. They literally pushed me out. I mean, I was flunking everything.

Every little child has the experience of being given a coloring book that has figures, and you're supposed to color the figures. Usually, the kid will say to the mommy, "Jeez, look, Mommy. I made this for you. Look how beautiful this is!" And the mother will look at it and say, "No, no, no, no, no. The sky should not be red and these people should not be green, and look where you've crayoned outside the lines."

Now she really thinks that she's doing something to help that child: "Oh, he doesn't quite get it yet. I'm going to help him to get it right." No, that's not what she's doing. She's destroying the ability later on to be creative, because I can give you a quote, "Coloring outside of the lines leads you to the slippery slope that eventually takes you to thinking outside the box." But if you've already crushed that in a kid when he's three years old, he's not going to have that later.

Stern: How do you define creativity?

Michelson: It's simple: creativity is when you can combine things in ways that other people generally would not think to do it. When you can solve problems in a nonlinear way. When you can find novel solutions to problems. I'm sure you've read on creativity and the concept of functional fixity. Almost all my inventions are in the medical field, and I can tell you that the medical field certainly has a surfeit of very, very bright engineers, because it pays well.

So when I would go to these various companies—and I've interacted with twenty of them commercially—there would be a room full of bright engineers. Now if you bring them something and say, "This isn't really doing what I want it to do." They reply, "Okay, what is it you want it to do?" And you tell them. Their approach to it is to add something on to it to make it do that which you want it to do. It's almost inconceivable to them that they should dismantle what you've now given them, atomize it, deconstruct it, and then rebuild it in a new way that has fewer parts, is simpler, and does all the things you want it to do—and not the things you don't want it to do. They do not have the self-permission to deconstruct.

Stern: Your field is in spinal orthopedic surgery. When you started out, did you immediately see the problems when you were in school, or was it at a point when you started doing surgery that you realized that there were physical problems in the instrumentation and the implants?

Michelson: I was doing a fellowship with a doctor who was considered one of the most knowledgeable spine surgeons in the world. I asked him a question, and he said, "I don't know the answer, but I'll tell you what not to do."

He started explaining all the things that didn't work. When he got done, I said, "Okay, so what should we do?"

And he replies, "Nothing, just leave that alone."

Now that bothered me a lot. That wasn't a good answer. I say, "How are we helping this patient?"

He said, "I don't know about that, but we're not going to paralyze him."

He started telling me horror stories about where other doctors who were literally world-famous surgeons had tried this or tried that, and always with bad consequences.

Stern: Do you feel that in those situations, the doctors didn't want to fail, rather than they didn't want to learn something?

Michelson: No, they didn't want to paralyze the patient. Their motivation in not doing something else was that they didn't know what would work safely.

Stern: When you see these problems, what is the process that you go through to brainstorm a possible solution or solutions? You're not trained as a biomechanical engineer, you're trained as a surgeon. So how do you see the problem, and how do you get to the solution?

Michelson: Well, that was an interesting statement that you made—that I'm not trained as a biomechanical engineer. I'm not going to argue with you, but I'm not sure that I know any biomechanical engineer that knows more about the biomechanics of a spine than I do. So I'm not sure what that statement means.

Stern: If you see a problem and you want a method to fix it, an engineer will say, for example: "Let's put a rod in there that's this long and has these certain qualities." They could specify the materials, they could draw something up, or they could make a prototype. My question is how do you get from point A to point B? Are you sketching? Do you give your ideas? Do you talk with machinists? I'm just trying to understand the process that you go through to come up with these things.

Michelson: I think it's generally held that Einstein was a brilliant mathematician. But if you actually read about his process, he did very simple mind experiments. After he got the insight, he struggled with the math. A couple times, he had to go back to one of his math professors and ask him for help. He knew he was right, he just didn't know how to mathematically prove it.

I think it's always a matter of imagination, not a matter of what you're talking about—the mechanics of it. So I think it always has to be an active imagination, with one exception—and it's a powerful exception. Sometimes you don't have to invent anything. All you have to do is look at things, but look at them through an eye of understanding. It's Ecclesiastes: "There is nothing new under the sun."

Stern: Once you see these problems, how do you communicate that to actually get the invention then?

Michelson: When I started out as a spine surgeon, we only did two operations. One was a decompression, where a piece of bone or a piece of disc is removed. The other thing we used to do was a weld job, where something's mechanically unstable. Your spine isn't a rod, it's an intercalated system of shock absorbers and blocks. So when the system goes bad, we take out the rubbery shock absorber and we weld together the two blocks. That's a fusion.

Those were the only two operations that spine surgeons did when I studied. So we either put people in body jackets or casts, put them in bed for six months, or put in some steel. We'd brace them together internally by using steel rods, or hooks, or plates, or screws, or something.

Now, along came things like pedicle screws, which were a potent form of internal stabilization, technically difficult to put in and requiring an even larger dissection. Adding pedicle screws, the fusion rates went up to about seventy-eight or eighty percent. Complication rates were higher than ever, and the complications were more severe. If you look at what we were doing, in essence, there were three separate procedures that took between five to eight hours of operation. So now you're asking me, how do you get the idea that there's something to be fixed? My first thought was, "Why don't we do them simultaneously?" If I said that to a spine surgeon, what do you think the response would be?

Stern: It's not possible?

Michelson: I think it would be the same as my father's: "What are you? An idiot?" I've heard that in different ways and forms my entire inventing career. It gets to point that when people say that, you know you've really got it.

I think I have close to a thousand patents and something like three hundred or four hundred US patents. When people say to me, "You know what, you didn't really invent that. I thought of that." You'll be standing somewhere and somebody will walk up to you and say, "You know that thing you think you invented? You didn't invent it. I invented that." You hear that all the time. Initially, I used to have some internal reaction to that, but I realized what they're really saying: "I've now looked at what you've done, and it's so simple, I should have been able to think of that." So that's my thought.

Stern: What do you think prevents people from doing it, if they have had the thought?

Michelson: It's only simple after the fact, once you've seen it.

Stern: What is your motivation to see these things, and go forward with them?

Michelson: So what motivates Gary? Despite probably being one of the most successful inventors commercially in terms of money, it was never money. In terms of aggrandizement, there's a long-venerated tradition in medicine to name anything you come up with after yourself. So we have DeBakey clamps. We have Cloward retractors. I defy anybody to find any single device that's ever been manufactured that has my name on it. And there are tens of thousands of my devices out there, estimated to generate $1.5 billion in sales a year. My name is not on anything.

So there's a reason for that. Even when I was a medical student attending a meeting, it became very apparent to me that you see these doctors who are pretty much one-trick ponies: they came up with this procedure or that procedure.

Here they were, and despite all the evidence to the contrary, they're still defending to the medical community that theirs is the best way to do it. I looked at that and said everybody likes to have an "Atta boy! Good job! Smart guy!" But I never want to be in the position where I look like one of these dinosaurs who's up in a meeting, defending what I should not be defending, because it may have been good at one time.

Just like I came up with something better than what existed before, somebody else has come up with something better still. I don't want to be invested. You know, the Zen Buddhists say everything you own owns you—and that's the same with inventions. You don't want to be the prisoner or compulsory caretaker of the legacy of your invention.

Stern: So what is the motivation?

Michelson: When you're a surgeon, they call it a practice because you're getting better and better. You're supposed to be a black belt in spinal surgery. Hopefully, you're still learning something. Each time you help a patient, you should have a sense of achievement—a sense of having done something worthwhile. You say to yourself, "If I'm struggling with this problem, I better go ask somebody and not take an egotistical position like I'm the emperor who can't admit I don't know how to do this."

So you go to your colleagues and you say, "I am struggling with this problem. What do you do?" They all say, "I don't know. I have no answer." Now you know that whatever answer you come up with, if it works, they need the answer as much as you do.

Stern: If the world always says "that's a crazy idea," how do you get the first person to say, "I'll try this thing."?

Michelson: I did not go out and try to convince doctors to do anything. I never stood up in front of a room full of doctors and said, "I did this, and it is great."

I never tried to be a salesperson. When I was doing spine surgery, we started putting plates on cervical fusions. That raised the fusion rate. That allowed a patient to get by without wearing a brace. But all the plates that I used had problems with them, and the problems were pretty obvious.

One day, a new plate came out. Sofamor Danek, which was one of the leading spine companies in the world at that time, made it. I wanted to talk to the engineer who was in charge of this plate. The guy gets on the phone, and I said to him, "Listen, I used your plate today. I have a list of its problems—and the solutions." I didn't just say, "Here are the problems." I said, "Here are the problems and here are the solutions." It seemed pretty easy to me.

The guy says to me, pretty much, "You're an idiot." He did not use those words because I was a doctor, but he pretty much said, "You're an idiot." He said, "Listen, you're only saying this because you have no idea what you're talking about. You're not an engineer and you don't know squat about squat. All these things are compromises. This is the best you can do because blah, blah, blah."

I say, "No, no, no." And he's getting annoyed and he finally says to me, "You know what, doctor? If you really think it's that easy, why don't you just do it yourself?" The reason that story is so interesting to me is that I had never in my entire life done that before. I had never taken something that already existed and thought, "I'm going to make a better one." I never did that.

Stern: What do you think are the skill sets needed for an inventor to be successful?

Michelson: Before we get to a skill set, let's get to personality or state of mind, or a way of thinking. You can't be rigid, where you're afraid of failing. Yogi Berra said, "You can observe a lot by just watching." You have to be able to see.

All the same things that Darwin observed were there for everybody else to see, but he saw them with a new eye of understanding. You have to give yourself permission to deconstruct things. I would say that's probably the number-one thing: you have to be able to take things apart. Like Einstein did, you can do it in your mind. You don't have to do it physically.

When you were saying, "You're not a biomechanical engineer." Well, in fact, one of the differences between me and most other physicians is that when I went to a company and showed them something, I was actually showing them the very thing they ended up making. It was already made. I had my own machine shop. It wasn't perfect. Usually, it was too complicated. But what happens is, if you allow the thing to talk to you, if you look at it with an open mind, it will teach you as you're using it.

Stern: Do you sketch these things out first or prototype?

Michelson: It really is the Einstein thing. You see it in your mind's eye. You've now taken apart all the pieces and sometimes you just have to throw it all away. This is the biggest struggle for people: overcoming this idea of functional fixity. Sometimes you can't get there from here. You can't leap a chasm in two small steps. So what you need to do is do something different. You just make a list of what you want. I want an instrument that simultaneously decompresses and fuses this segment. What do I not want? I don't want to injure the blood vessels. I don't want to injure the nerves. I want to minimize the blood loss. So you make up this list, and then you say, "How do I do this? What's the fewest number of pieces needed to do this?"

Stern: Have you had any failures?

Michelson: There's that Edison quote: "I have not failed. I've just found ten thousand ways that won't work." I don't remember having any kind of monumental failure. I had things that didn't do what I expected them to do, so I learned from that. It's an evolving education. You don't say you failed, because you're improving what you're doing. When I give something to somebody, I don't want him or her to be able to say, "Here's what you need to do to make it better." If they're right, they're right, but hopefully, I did that already.

Stern: Further to people giving you advice, have you had mentors? If so, what do you look for from them? And are you now a mentor?

Michelson: Let me answer the second one first. Whenever somebody contacts my business office and says, "I'm working on something. Do you think the doctor would talk to me?" I talk to everybody. I give them as much help as I possibly can. Now I'm involved in a lot of medical research because I have a big foundation. That's what we do, and I interact with researchers all the time.

Now, as far as getting inspiration. I never had somebody say to me, "That sounds like a great idea—go ahead and do it!" The only thing I ever got was the same one I got from my father: "You're an idiot."

I never got any positive, "Gee, that sounds like a great idea!" I don't remember having that. And I can tell you this about my first really successful invention, which is the thing I told you about where you get to do a decompression/fusion instrumentation as one procedure that takes forty-five minutes. The world's most preeminent spine surgeon inventor was a guy named Ralph Cloward. He invented things in the fifties that we were all doing from then up until the eighties, with nobody supplanting him, nobody coming up with a better way. So the things he had invented persisted for thirty years uncontested. And, of course, he was like a god. So when I was doing my spine fellowship and I was in practice, every time I had a chance to talk to him, meet with him, I always did.

When I came up with my first invention breakthrough, I put it all in a box—the instruments, the device, everything—and I sent it to him. I said, "Ralph, can you look at this? I need it back as soon as possible." A day goes by, weeks go by, and a month goes by. Finally, I pick up the phone, I get a hold of his secretary, and I said, "Can I talk to Dr. Cloward?" And he gets on the phone. I said, "So, Ralph, what did you think?"

And he replies, "About what?"

I say, "About my stuff. Did you look at it?"

He replies, "Yeah. Don't be an idiot." These are the words he said to me and I'll never forget this: "What's goddamn wrong with you people? Why won't people just do the operation exactly the way I've taught them?"

So here's this guy telling me, "Don't be an idiot! Just do what I taught you."

Stern: What advice do you have for inventors out there then?

Michelson: Well, first of all, I've got a friend who's a psychiatrist—he's not my psychiatrist, he's my friend—and he once said something. I responded, "Hey, that's great! That's really great!"

And he said, "Don't do that."

I asked, "Don't do what?"

He said, "I don't want your praise."

I said, "Well, wait a second. It's not flattery—you actually did something. It's very neat and I was just saying to you, 'Great job!'"

He says, "Yeah, but I don't want that, because if you then think I did a bad job, I don't want to hear from you, 'Bad job!' I don't want you deciding for me if something is good or bad."

He was expressing it forthrightly, but it's correct. You have to decide inside yourself what's right and wrong. You have to believe in yourself. Don't expect the rest of the world to believe in you. If you're looking for them to tell you that—"Yes, this is a good idea. Go do it!"—you're wasting your time.

Stern: Do you have any inventions out there or products that you like?

Michelson: I think digital cameras are incredible—the fact that you can have a camera in your phone and can take pictures of whatever you want. I think the fact that there's no teenager in the world wearing a wristwatch because they can't understand what the hell those things are for since the time is everywhere. I mean we're living in a world that is changing at an accelerating rate.

Stern: What do you do for fun when you're not doing your work?

Michelson: I have three foundations, and I get great satisfaction out of each of those. We have a foundation that funds cutting-edge medical research that usually the NIH [National Institutes of Health] has not come to fund yet.

I have another foundation that's an animal welfare foundation. We have thirty-five full-time people. We try to alleviate the issue that the municipal animal shelters of the United States kill about four million cats and dogs every year. They're not sick, and they call it euthanasia, which is lie. Euthanasia relieves suffering. These animals aren't suffering. They need homes.

I have a very large research program out right now to find a single-dose chemical sterilant that you could inject in a cat or dog to sterilize them, rather than having to do spay and neuter surgeries. That is the holy grail of this program. We put up a $25 million prize for anybody who can produce it, and we have committed $50 million to fund the research to help win the prize. We've got twenty research projects ongoing and we're looking for more.

Then I have a foundation that's called the Twenty Million Minds Foundation. Our goal is to produce all of the textbooks that are used in the college space—both the community college and four-year college levels—but are better content-wise than the ones that come from the large publishers. They are digitally available, where they're hyperlinked. So you read something and then it asks you some questions. If you did not get the answer right, you can't get it wrong, it then takes you to an enriched reference source to help you understand what you didn't get.

But the point is, textbooks like this have never existed. We're making it where you can have it in paperback or hardback if you want—but you can go online, where it's all free.

Stern: I actually asked what you do for fun.

Michelson: It is fun. That makes me feel good.

Stern: Do you plan to retire? And if so, what will you do?

Michelson: Well, I retired from spine surgery in 2001, so I don't operate and I don't try to invent in the area of spine. What I do is actively manage these foundations. It's not like I passively fund them. I interact with the researchers and sometimes I help shape and steer the research that's being done. I actively run the animal foundation and the Twenty Million Minds Foundation. We have board meetings every week.

Stern: Any advice you want to offer to an inventor?

Michelson: Everybody's different, but everybody's got problems. If you're struggling with a problem in your ordinary life—it doesn't matter what, it could be your kid keeps spilling the Cheerios on the floor—so is the rest of the world. Your advantage is that you identify this as a problem and then you can ask how you solve this problem. Now, some people honestly don't have the imagination or the technical ability, and they're not going to do it. But other people will. So it all begins with one of two things, either asking the question or else simply observing something and seeing it in a different way than everybody else did, and say, "Here's what I can use that for, and here's what I can do with that."

Stern: Do you have a definition of what invention is?

Michelson: I think the patent office came up with a pretty good one. It has to be novel, which means it wasn't preexisting, because if it was preexisting you're not the inventor. And it's supposed to be useful.

Stern: Do you have a definition for innovation?

Michelson: Innovation is creating something that did not exist before, and that's not obvious.

Stern: You've offered a lot of quotes, and I just pulled up one called Planck's dictum. Max Planck won a Nobel Prize in physics. He wrote: "A new scientific truth does not triumph by convincing its opponents and making them see the light, but rather because its opponents eventually die, and a new generation grows up that is familiar with it."

Michelson: Yeah. In medicine, it happens a little differently. What happens is, the first time you show it to them, they say you're an idiot, they tell you you're wrong when you show them the evidence. But a year or two later, if they can put their wagon to that star, you'll see them up there at one of these big meetings, and they're giving a talk on how they do this and how they do that, it's based on this and that—and it's like they own it, they came up with it themselves. They don't even remember that they told you no.

Stern: You're not going to retire, so I'm assuming you're just going to keep on working?

Michelson: Yeah, but I'm not going to keep on inventing medical devices. You're never too old to reinvent yourself.

How many of these interviews have you done already?

Stern: You are number seventeen.

Michelson: I've got to ask you. Did I sound a little bit more whacky than most of these people?

Stern: Generally, the majority of the people I've been talking to are from the corporate side. Their view is a little more—and I don't mean this in a negative way—focused. They talk the corporate talk. You talk—and I do appreciate this—you talk more in stories and allegories, and that's really fun. I think it's all about telling a story. It's a positive form of communication. You have actually taken the whole commercialization side out of it. It's not a motivation for you at all.

And, frankly, that's refreshing, because on the corporate side, there is that commercial awareness continuously—every single moment there's an awareness of that. In a corporation, the remuneration may not be direct, but you're getting your salary or you're working your way up the corporate ladder. So, yes, you are more flamboyant. I would say when I'd ask questions about creative process, the corporate person tends to give me a delineated direction for step A to step B. But there's no right way or wrong way, so I'm just trying to understand what you're doing and how you're interpreting it. There's no judgment on my side.

Michelson: No, I wasn't worried about that. I am in the Inventors Hall of Fame, and I've interacted with a lot of these people. It's rare to find somebody who has—I don't know—self-analyzed themselves psychologically to really get back to the roots of this thing. If they had, most people would treat that as something they don't want to tell anybody. You know, they're not going to say, "Here's what was really going on."

It seems like you're laying yourself bare. In a lot of ways, I think it's a little bit like an artist, where the artist puts himself out in the work and really kind of wants you to see it. I wasn't worried about it being right or wrong. I just thought it might be a very different interview than anybody else has ever given you.

Stern: Yes, it was. It had a lot more energy, a lot more enthusiasm. You weren't afraid to say what you were feeling, and I greatly appreciate that. I would say that you offered the emotional content, which I really didn't get from anybody else.

I think most people, certainly in the corporate field, are much more reserved. I think you let it all hang out, so I appreciate that.

Michelson: You know the great quote by the industrial designer Raymond Loewy? Somebody accused him of being a stylist posing as an engineer.

Stern: He was not shy guy.

Michelson: He said: "I sought excitement and taking chances. I was all ready to fail in order to achieve something large." You know, when there's a certain joy to what you do, and when you have this certain energy, that's what makes you happy. Doing those things made me happy. That was the real motivation: it made me happy.

Al Maurice

Global Fellow
The Dow Chemical Company

Al Maurice joined the Rohm and Hass Company (which is now a part of the Dow Chemical Company) in 1982 after completing his PhD in physical chemistry at the University of Illinois. For the next 17 years, he worked in the emulsion polymer synthesis field, holding positions of increasing responsibility. His primary focus has been architectural coatings, with excursions into traffic paint, ultraviolet-cured coatings, and leather chemicals. In 1992, Maurice was promoted to synthesis section manager for architectural coatings. He became a project leader in 1999, with responsibility for coatings synthesis and applications research. In 2004, he took on the role of global technology manager with responsibility for global segments of architectural coatings research and development. In 2010, he moved up the technical ladder, becoming a Dow Fellow.

Throughout his 30 years with Dow, Maurice and his teams have consistently delivered commercially successful products. These products include acrylics, vinyl acrylics, and styrene acrylics. He has a strong technical background in emulsion polymerization, a wealth of experience in architectural coatings and a practical approach to industrial research. He is an inventor on numerous patents and a dedicated mentor to many. In 2002, Maurice was the recipient of a Vice President's Award for the investigation and delivery of alternative supply sources for key raw materials resulting in significant cost savings to the company.

Brett Stern: Please tell me about your background. Where were you were born? What was your education and your field of study?

Al Maurice: I was born and raised in Chicago, Illinois, so I'm a Midwest boy at heart. I went to Eastern Illinois University as an undergraduate majoring in chemistry with a minor in math. Then I went on to the University of Illinois, where I got my master's and a PhD in physical chemistry. After that, I joined Dow.

Stern: Growing up, would you say you were an inventive or inquisitive kid?

Maurice: Well, I could say I was. I always liked to understand everything about everything, which is what forms the basis I think for the inventive process. If you don't know what the current state of knowledge is, it's pretty hard to invent something new. I was always a good student, with a natural bent toward math and science. It wasn't until I really got into college that I decided to go into chemistry. I always liked math a lot, but I viewed math as being a tool that enables you to do other things. It's not an end in itself so much. I know that for some people that's clearly their career and they want to become mathematicians. But for me it was more of this is a great tool and you can use it to calculate and figure out other things.

Stern: Could you give me some background in technical or layperson terms about the type of areas that you work in at Dow?

Maurice: For most of my career, I have worked in the coatings area. Dow does not make paint but what we do make is many of the components that go into a paint. People say paint is paint, but paint is a pretty complex composite of binders, polymers, thickeners, pigments, extenders, and dispersants. It's a pretty complex mixture, and what Dow does is develop the components that can then be used by paint companies to make paint.

Stern: So your work has been in protective coatings or improving protective coatings for homes and walls?

Maurice: Exactly. And if you think about the paint industry—and it's primarily house paint that I work with—back in the fifties and in the sixties, most paint was solvent-based. If you're my age, you'll remember going down in the basement and seeing all these hardened, bent up brushes sitting in coffee cans filled with turpentine. What we have done is essentially help to convert much of the architectural and even a lot of the industrial market over to water-based paint. It required a very fundamental change in terms of the components that we're dealing with. Because now instead of being a solution in some kind of linseed or whatever solvent, you now have the system in water.

So instead of having solution polymers,[1] you actually have polymer dispersions, which are basically 100 nanometer-, 500 nanometer-sized polymer particles that are suspended in water. You need to be able to take these particles and have them flow out and dry into a nice, smooth film that gives you the protection, the beauty that really makes your house look nice.

---

[1] Polymer: Composed of repeating structural linked units or chains typically connected by covalent chemical bonds each a relatively light and simple molecule.

Stern: When you first started, you were doing the actual chemistry of this, correct? Would you say right from the beginning you were inventing things? Or was it in a sense just redoing something?

Maurice: Oh, no. I would say at the very beginning, you are just trying to learn how to do polymerizations. But for the bulk of my career, I have spent time actually doing polymerizations, or instructing or working with other people who are running polymerizations. [We worked to] further the boundaries and push performance to new areas—whether it is new properties or whether it is improvements upon existing properties. Properties that we've been able to improve include stain blocking, washability, adhesion to a variety of substrates, film opacity, odor in the paint, and flow and leveling.

Stern: The product that you're selling, in a sense, is not an end product. You were fairly removed from the actual end user. How did you find the problems to work on then? Does the customer come to you? Did you go into the market place?

Maurice: We talk about it in one of two categories. First, there is a technology push, where you're working in the lab and you come up with some new chemistry, some new reaction, and some new component that is going to give you some improvements in performance. Then what you do is you can take that technology and build it into something. Then you go to your customer and say, "You haven't asked for this, but this does something that we think you should be really interested in."

Stern: In a sense, you actually had the solution but you didn't know what the problem was?

Maurice: We have been in the coatings business for a long time. We have a fair amount of experience within the company, both from a technical position and a marketing position. We have a pretty good idea [about which problems to solve]. We stay pretty well attuned. For example, tomorrow I actually have a meeting with a paint company. Last week I was out with another paint company. So we have fairly frequent interactions with customers. We're not just throwing something over the wall and saying, "Well, I have no idea what this is good for—but, here try this out."

Actually, I should probably step back because I'm getting ahead of myself a little bit. Most of my career has been involved with the synthesis of the polymers. That means we make the polymers. But the other half of our team is in applications, which means they will take and make paint out of those polymers and test those paints.

We'll go back and forth, and we will work it and optimize it. This is where we can go to our customers and say, "Here is a polymer. If you formulate it this way, you can make this kind of paint but, you don't have to formulate it this way. You can formulate it another way, but this just gives you an example of the type of performance you can get with this type of technology that we would like to offer to you."

So you show them a value proposition: "Oh look, this has really good scrub resistance, or has really good wash ability, or has really good exterior durability. This is why you should want to buy this polymer from us because it is going to be better than what you can get from anyone else." So we're not just throwing things over the fence to say, "Hey, this might be good for something. Try it out."

Stern: Does the opposite happen? Do the companies you are working with come to you and say, "This is the problem. Can you solve it?"

Maurice: Yes. And I'm sorry I made a long story out of this, but the second half of new project ideas is when the customer comes to us and says, "You know, we'd really like a paint that has better stain-blocking or has better color retention when we put it outside. We have something that lasts two years. We want something that lasts five years." There are a myriad of different things customers ask for.

Also, there is a continual issuance of government regulations, like, "We don't really want you to be using alkylphenol ethoxylate surfactants anymore." Or, "the VOC [volatile organic compounds] of your coatings must be less than 150 grams per liter," or those kind of things. You see those things coming down the road from a regulatory point of view or from consumer advocacy groups, and that can influence the dialogue. Our customers will come to us and say, "Well, we've been buying your product X for a number of years, but we can't buy this anymore because it doesn't meet the new regulations such as Proposition 65 in California."

Stern: Is it easier for you to develop the solution or for the company to come to you with the problems?

Maurice: I would say it's different going either way. When you're doing the technology, you build a point of view. You have a little bit more freedom and then maybe a little bit more time, but then sometimes internally you have to justify that effort more. Whereas if a major customer comes and says, "I want a new polymer that does this," that's pretty easy to get a project approved because there is a clear route to market where someone is willing to pay for it. But a lot of times you really need both. And another aspect of it is that sometimes customers will ask you for what they think they need, which may not actually be exactly what they do need.

Stern: What do you do then?

Maurice: Customers are usually not so prescriptive that they say, "Make me this exact thing." They know that we have more expertise in terms of making polymers, and they can suggest the direction you take, but you can say, "Well, we looked at that but it didn't really work so well. But here's something that actually does move you in the direction you're asking for."

Stern: So you work in a corporate setting. Can you tell me the process that you have to go through in your daily routine to do your work?

Maurice: It has changed over the years as I have taken on different roles. Currently, I'm an R&D fellow. I have a little bit more freedom than a lot of people, so it's actually a really, really cool job.

Stern: What's cool about the job?

Maurice: Well, where I am right now, I have very little micromanagement, which is a style that I prefer. I like to say to someone, "Tell me what you need. Tell me the goal. I'll figure out how to get there."

Don't tell me to "step to the left, then one foot over here, and then turn right." For example, if I'm driving from here to Chicago, tell me that you'd like to go to Chicago and I'll figure out how to get there. Don't tell me which highways.

So part of it is that, but the other big part of it that I really like is that I get a chance to meet with customers. I get a chance to do a lot of mentoring of younger people. I get to interact with people in different functions and businesses and I also get to be involved with basically all the projects that are ongoing in architectural coatings.

When you're on specific projects as say a senior scientist or something like that, sometimes you have responsibility for a portion of the total portfolio as opposed to the whole thing. I prefer to have the broader picture of seeing how everything fits together, how the different components combine, and the chance to look for gaps in the portfolio. Because sometimes it's like, "Oh, these guys are working on this thing over here, and these people are working on this thing over there, but if we put those together, we can make something better." Whereas if you're working on *these* three projects, but you're not involved with the five projects *over there*, then it's hard to know what everyone is doing or how to cross-fertilize ideas.

Stern: You have been both a scientist and a manager in your career. It sounds like you enjoy the management side to get the bigger picture.

Maurice: Well, I've enjoyed both sides. If you ask me right now, I prefer the technical side to management. Depending on your group, when you're on the management side, you may only have a portion of the whole. There were several managers in architectural coatings, for example. I like the ability to influence a broader range of projects, programs, and people.

Stern: Does Dow have a particular method of encouraging people who want to be scientists or who want to be managers? Do they allow people to go up the corporate ladder and to grow in either profession?

Maurice: Yes, I would say they do. Some people are very one side or the other: this person is clearly a manager, this person is clearly technical. I was a little bit of both.

I was a manager for about fifteen or sixteen years. Now I'm back on the technical side. If you ask me which one I prefer, I prefer to be on the technical side. A technical role also has less administrative responsibilities.

Stern: Would you say that being a manager made you a better technical person? And if so, how?

Maurice: Actually, that's a very insightful comment. I think so. At least, I hope so. The reason is that there are a number of people who are on the technical ladder and on the manager ladder. Some of the people who are on the technical ladder have been there for their entire career. They are very smart people, and they are inclined to be exploratory and creative and innovative or inventors.

But on the other hand, I think being in a managerial position where you need to deliver something at the end of the day forces you to take a more pragmatic or structured approach. I think that has given me a little bit more focus and a little bit more of an ability to say, "Yes, there are lots of things to explore, and there's lots of things that you can do as a technical expert. But some of it has more impact on the company, and on people, and on customers, and on product development than others." It helps you to prioritize. I think that I may be in a better position to utilize my past experience to help my current job as a technical expert.

Stern: Would you say there is any particular invention that is your favorite? One that you've worked on over the years?

Maurice: Dow is very big on acrylic chemistry. A number of years ago, we wanted to do some things in a different type of chemistry. I was given the opportunity to lead an effort on vinyl-acrylic product development. I started out with a little bit of time. This is perhaps an example of how the innovative process works. It's a case where it is a similar chemistry—but it's not all-acrylic chemistry. Incorporation of vinyl acetate makes it different. The rules on how they polymerize are different. The things you need to do to make really good vinyl acetate polymers are not necessarily the same things as you need to make good acrylic polymers. It's a new chemistry, so you have to learn the new rules.

I got a chance to do some of that and we ended up with a high-performance vinyl acetate product. We started from a blank piece of paper and spent some time to really learn the rules. And then we worked at changing it and looking at different variables, because there are a number of synthetic variables that go into making a polymer.

We did a lot of evaluating the rules and we were able to put together something in a pretty short period of time that ended up looking very, very good in terms of being competitive with products that were on the market—and then some. Once we had some success with that, we went beyond it and said, "What other new properties could we now build in?" We were able to do a number of variants that brought other performance attributes such as improved stability, higher solids, and excellent alkyd adhesion.

What you get out of this type of work is a lot of satisfaction and pride, because we started from scratch. We built this thing the way we thought it should be built, and then we took it to the plant and scaled it up. And it ran in the plant just like it ran in the lab.

Stern: How long did those projects take?

Maurice: It took a couple years because we had a number of different projects we were working on at the same time. So the first generation maybe took two years and then we banged out a couple of other variants with different performance attributes over a shorter time period. Sometimes we would do a bit of a launch and learn approach where some customers liked what we had, which is great, but sometimes customers said, "Well, that's really great, but if you had a little more of something else, I'd really like that." Then you say, "Okay, I think we can do that."

Stern: It sounds like when you are working on these projects, you are doing it in a group. Correct?

Maurice: Absolutely. This gets back to the comment about working in a corporation. When you're in grad school, you work by yourself. If you don't do it, it doesn't get done. But in a corporate setting where we're in R&D, synthesis and applications work together. One makes the polymers and one makes the paints from those polymers. Then we also have a scale-up group that we work with. We work with the sales and tech service groups that are more closely associated with customers. We work with our marketing guys. We work with our legal guys for IP and things like that. So to quote Hilary Clinton, "It takes a village." That's why I throw out a lot of "we's." Because it's not an "I did this, and I did this, and I did this."

If it were just me, none of the products that I have developed over the years would have ever seen their way out of the lab. Because I'm not going to take it and make a paint out of it, and then run to the customer with it, and then run to the plant. One person can't do everything. I'm not doing that.

Stern: In doing these collaborations, what happens when you get to the point where you have two directions or two possibilities? How do you decide which is a more effective way to go?

Maurice: If they are competing technologies, you play them both out until you determine if there's a clear advantage for one versus the other.

Stern: So it is somewhat democratic in deciding which direction to pick?

Maurice: I would say not fully democratic, but it's mostly democratic. We have team meetings, we discuss things. We use a stage-gate process for development. That's fairly common for a lot of product development. Basically, most R&D programs are done through a stage-gate, so what you do is have multiple stages, and depending on the system, it is typically about five gates or so [that you must pass through].

The first stage is where you are defining an idea, and then you shape it, and then you develop prototypes, and then you scale it up, and then you commercialize it. There are multiple steps and there's a series of activities that are associated with each stage. You know when you're developing a prototype that you need to spend a lot of time trying to get something that matches the objectives for the program.

For the shaping part, you might talk to different customers to try to figure out what the reasonable objectives might be. And so there's a predetermined set of activities associated with each stage. Before you move from one stage to the next, you have to go through a gate—and the gate is typically overseen by several stakeholders who will say, "Oh, great job, team. We love what you're doing. Keep going that way."

And, "Is there anything else you need?" Such as more resources or more whatever. Or you might get, "You guys can't decide between going left or right, so we'll make that decision for you." Or you might need capital or something—or whatever is involved. But basically, the team will move the program through those stages. At each of the gates, it's time to tell the stakeholders where you are, what you've done, and what you'd like back. If they agree, then you move to the next stage. And if they don't agree, then you don't since stakeholders can also terminate programs.

Stern: You mentioned that you work on multiple projects simultaneously. Is there an advantage to that?

Maurice: Yes. In the typical workflow, you're making polymers and then you're making paints. So when you're working on a development, the idea is that you make a series of polymers and use a design of experiments or whatever approach you're using to cover a certain area. And you check to be sure that you are moving in the right direction. When you give them to the application guys, they're going to have to make paint and then test the paint. It's hard to start the next series of polymers because you haven't gotten the results back on the last series. So for example, you start working on a second project while the polymers from the first ones are being evaluated.

Stern: How do you start a project? What are the tools that you use to develop an idea, to ideate a solution, to brainstorm? Do you sketch, do you prototype, do you model?

Maurice: We use a little bit of a lot of different things. Typically, you get the team together, and at the start, you make sure you understand the goals. This is the time to question the goals and to make sure they make sense. You ask questions like, "Do we really need this? Is this a must-have or is it only a like-to-have?"

Sometimes people will say, "I've got this idea," and a lot of the things that we do will build off of our existing knowledge base. Typically, what you'll do is say, "There's a polymer that we have that might be a good starting place, but we have to modify it to do this, this, this, and this with it." From there you may want to benchmark the current state of the market. For example, you might say, "I'm doing a new paint and primer project, so I'm going to the stores to buy all the paint and primer products I can find. Then we'll evaluate them and see what they look like, and find out what's good about them and what's missing."

Stern: I have to ask you: do you actually watch the paint dry?

Maurice: [Laughter.] No. The answer is not so much that we sit there and watch it dry, but we do a lot of evaluations of the paint after it has dried. We'll dry it under different conditions. Does it crack if it's cool outside? If you treat it with rain fairly soon after you apply it, is it going to suffer and get blisters or something? Although the drying environment is extremely important, especially for low VOC paints, but we don't typically sit around and watch the paint dry. It just dries by itself.

Stern: It was a good visual though.

Maurice: Yeah, that question has been raised many times.

Stern: In actual developmental work, are you mixing things together? Is it very hands-on? Or is it more theoretical?

Maurice: Well, if you're in the lab, it's very hands-on. I do not do a lot of lab work anymore. I was in the lab for the first ten years of my career, and at the time I thought—[verbal sigh]—I was ready to move on to something else. But in hindsight, being in the lab for an extended time period is really an advantage because I've seen a lot of the things firsthand, not just hearing people talk about it. I've done things. I know how to do polymerizations, and how we do the polymerizations, and what things should look like and what they shouldn't look like.

It helps me now a lot because I sit down with people to go through what they're doing and how they're doing it. Because of my past experience, I know exactly they're talking about and the potential things to avoid.

I also continue to work with our lab in China pretty closely. It's great working with the people in the lab in Shanghai. But it's one of these things where you work with different people and you make assumptions about what they are doing or how they are doing something. If your assumptions are incorrect because you didn't discuss the details, you can run astray. Based on experience, as well as depending on who you're talking to in the different labs, or even different people here in the US—some people are much more in tune with the details than others. It is often the little things that matter, so you need to ensure you understand exactly what is being done.

Stern: What skill set do you think an inventor needs to be successful?

Maurice: That's a *really* good question. I don't think there's any one answer for this, but I would say that first of all, you need a knowledge base to work from. It's hard to build if you haven't got the foundation. The second thing is I think it's important to have an open mind, and to listen to what people have to say, and to really make sure you understand the problem before you try to come up with the solution. It's important to talk to different people because different people have different perspectives and different knowledge, and a lot of times people will have pieces of the puzzle, but by interacting with different people, you can maybe put it together.

In fact, a lot of times where I personally will get information or ideas [is in conversation.] For example, a colleague at work may say, "We're doing a new chemistry for this," and you say, "Oh, that's kind of interesting. We're not doing that in coatings. Maybe we should be doing that in coatings, too." Or they're telling you about a new functionality or a new monomer, or a new type of polymerization, and you're like, "Gee," or even "wow." The chemistry specifically may not work for you, but the mechanism may be something that you can apply.

I know that if I go to a seminar or a conference to hear someone talk, I'll often drift off because the mechanism, or the chemistry, or the approach that they're talking about spurs an idea that might be useful in solving or addressing some of the problems I have. Even when it's not really what the guy or lady is talking about specifically but something analogous to that.

Stern: Where do you seek inspiration or solutions? When you're staring out the window or watching paint dry—where do you seek inspiration?

Maurice: Sometimes it's reading journals. I try to keep up on different journals to find out what's going on, but sometimes it'll even be on TV or in a regular magazine. But there's no "Oh, I'm in my thinking spot, so now I'll think deep thoughts." That's not usually the way it occurs. Usually there's some kind of trigger that makes you say, "Aha, that's an idea."

Sometimes if you're truly in an exploratory mode, you want to be in a nice quiet room and just think deep thoughts, but you don't usually have time to do much of that. It's more likely that it's a problem-solving situation where you're like, "Okay, I really want to get a better stabilizer for my latexes because when I mix them, it coagulates" or whatever. Then what you'll do is think about the current system and ask what you have. "We've got these latex particles, which are just basically spheres of polymer, and then we've got the surfactants that are associated with them. So why isn't the surfactant keeping them apart? Well, maybe the way the surfactant associates with the surface isn't right." Is the surfactant like a lunar lander where it's got multiple association points, or is the stabilizing group too flexible? Do I need something stiffer?" And then you start thinking, "What could I use instead of the current model?"

I think what part of it is you have to delve into the mechanism of what's going on, because when you're trying to solve a specific problem, you want to think about what is going on in your system, knowing the components that you put in there. Then what you'll do is use a scientific approach, where you say, "I think maybe this is what's going on. What experiments could I do to test that?" Then you'll try some experiments, and from the results of those you'll say, "Oh, this seems to be getting better if I do it this way." Ultimately, we work to combine the wealth of past knowledge with new ideas to meet current and future needs.

Stern: You do these tests and get successes. What about failures? Have there been any really big failures? What did you learn from them?

Maurice: What kind of failures do we have? Basically, it could be unsuccessful projects. It's where you went along and either you never solved the problem or you solve the problem but the customer has said, "Well, that's nice, but that problem wasn't really that important for me." Or it's a case of, "Yes, but this requires me to do some other things that I'm not really interested in doing. So forget it." I would say, in general, it is a case where either you haven't fully delivered all the needs or else you perhaps misjudged the importance of certain aspects.

Stern: You said now you are in position to mentor people. How do you go about that?

Maurice: I like meeting with chemists and talking to them about their work, and hearing what they're doing, and what's working and what's not working. What I try to do for them is to not manage them. I try not to be a judge or an assessor. I do try to give them a little different perspective since I am not as intimately involved in their work. It is a bit of the forest through the trees type of thing. The other thing I probably do more of is hear what they're doing because people here do a really good job. The people at Dow are really, really good. So I'm more likely to say, "You're doing these four experiments. Here are a couple other things that might help you further your performance or move your project forward. You might try looking at these."

Stern: Does Dow have a formal system for having people mentor others?

Maurice: There has been encouragement of that for a long time. The problem is sometimes I think mentoring works better when people want to be mentored as opposed to telling people "you will be mentored." What I try to do is encourage people to share what they're doing, not only with me but with everybody. For example, I've set up a seminar series where all the synthesis people give a talk. It's billed as a presentation but it's meant mainly to be a discussion. They give the background on what they're doing. Others then respond with, "Have you tried testing this? Have you thought about that?" So it's a learning experience on both sides.

The other thing is what I call "functional excellence." The idea is to make sure that we have the best polymer synthesis group we can. From a safety point of view, and also from a knowledge base and a procedural and scientific basis, I put out a tech digest once a quarter. It's an opportunity for people to say, "I was working on this project and I found something I think is interesting." The idea is that it's not a full report on any one topic, but it contains a bunch of submissions from a variety of different people. If you're interested or you want to learn more about Joe-down-the-hall's discovery, then you should go and talk to him.

Stern: Do you have any favorite inventions or an inventor that you admire?

Maurice: I would say some of the people who worked with the company who were leaders in the polymer business when it first started. It was a bit of a gamble for the company because all paints were solvent based in that time period. Those guys entered the water-based paint market—and some of the early water-based paints were not very good—but they were at least in water and you could use soap and water to clean your brushes. But those guys really helped transform the market.

Stern: Is there a project that you're working on now or in the future that you could talk about that has you excited?

Maurice: You probably have heard about a lot about Evoque Pre-Composite Polymer Technology, a coatings chemistry that Dow has been promoting lately. It is a technology that is designed to improve the spacing of TiO_2 in the film.[2] The technology promotes improved opacity or hiding. For example, paints based on Evoque Technology facilitate greater opacity. So if you have dark red on your wall and you want to paint it white, you don't end up with pink. You really end up with white. Improving the hiding efficiency of TiO_2 is important since TiO_2 is currently in short supply and Evoque Technology allows you to use less TiO_2.

Stern: Outside of the paint field, do you have a favorite product that you just love having around your house or your person?

Maurice: I remember when they came out with the digital camera. I really like the fact that you can take lots of lots of pictures even if you throw most of them away because that is often how you learn. Also, you can optimize and customize the pictures when you develop them so that it looks like you're a better photographer than you actually are.

Stern: Do you have a favorite color? What color is your house painted?

Maurice: I'm colorblind, so that's one of the reasons I have to stay on the synthesis side rather than the application side. Actually, my favorite color is blue, but I really like orange and red as well. My first car was orange. At the time, I was dating my wife and I remember telling her, "I just bought a new car—and it's orange." She said, "What? You got an orange car?!" But I really liked that car and I really liked the orange color on it.

Stern: So do you have any advice for inventors out there?

---

[2]Evoque Pre-Composite Polymer Technology improves the particle distribution and light-scattering efficiency of TiO_2 while improving or maintaining hiding qualities.

Maurice: Anyone can have a good idea. It's not as if only the really smart people have good ideas. An idea can come from anyone and from anywhere, and a lot of times it's based on making connections between things.

The other thing to know is that inventions and innovations often take a fair amount of time and a fair amount of effort. The easier ones are when you have a great idea and everybody loves it. Yet a lot of times you have an idea that could change things, but it requires people to do things differently or think differently. So there's a curing time, if you will, on the idea. Don't be discouraged if people don't immediately embrace your idea. In fact, some of the biggest innovations have taken years to implement. Say for example, you come up with a new way of making polymers or a new polymer type that is going to have great performance. If it doesn't fit exactly into the current paint company's way of making paint, they may not immediately jump on that because they'd have to make changes, or they have to test it out. Sometimes you need a little bit of patience and a little bit of persistence to push through some ideas.

Stern: You said "invention and innovation." Do you have a definition of what innovation is? Is there a difference between that and invention?

Maurice: Yeah, that's a really good question as well. Inventions and innovations are not the same. Inventions are nice, but in the corporate world, it is extremely important to be able to bring innovations to the market in the form of new technologies and products.

My definition would be that inventions are something new. An innovation is more something that's new and is useful or can be converted into something useful. So you look at a typical company and you look at the patents that companies get. A lot of times you say, "Well, that shows how creative that company is." But just because you have more inventions and more patents doesn't mean you have more innovation.

Ideas and innovations can come from anyone or anywhere. People need to be open to new ideas and actively seek out a diversity of opinions and perspectives. To me, you've got to convert an idea into a product or something that is practical. Something that is useful and that people want to have. You need a balance of "pie in the sky" exploratory research with more practically directed research, which typically pays the bills. That's innovation. Sometimes I over simplify concepts, but it helps me make sense of the world. For me, when you boil it down, innovation is about understanding the universe and problem solving. In fact, that is the essence of my job as a scientist.

Time is also a critical element of innovation. It often takes time to come up with new concepts, and even more time to implement. Often the more novel the idea, the more time that will be required since it necessitates customers doing things differently.

Stern: So it's the commercialization of the invention.

Maurice: Yes.

Stern: What do you do for fun when you're not working?

Maurice: I cycle. And I garden. A year and a half ago I got a new road bike. There's some pretty clever gadgets that you can get for bikes. For several years now in September, several friends and I will ride our bikes from Pennsylvania across the Delaware River, across New Jersey to the shore. We'll stay overnight and then we'll ride back the next day. That's about one hundred fifty miles total.

Stern: Do you plan to retire? What will you do?

Maurice: Yes, I suppose someday. Sleep later possibly. Do more gardening and more cycling because right now I have time constraints on those things. I will also likely travel more. Actually, that's one of the things that I would also say has been a real nice thing about working with the company. I get a chance to travel. Usually it's to see customers or to meet with scientists in our lab in China or our labs in Europe. I think having the ability to go places like that is a valuable experience because it promotes sharing ideas and gives you additional sources of information and perspective about what people are doing and why they're doing it and how they're doing it.

Helen Greiner

Roboticist

CyPhy Works

Helen Greiner, born in 1967, is CEO of CyPhy Works, Inc., a start-up company that acts as a "skunkworks" to design and deliver innovative robots. In 1990, she co-founded iRobot, which has become the global leader of mobile robots from the success of the Roomba vacuum cleaning robot, and the PackBot and SUGV military robots. Greiner served as the president of iRobot until 2004 and as chairman until October 2008. Specifically, she developed the strategy for and led iRobot's entry into the military marketplace. She created a culture of practical innovation and delivery that led to the creation and deployment of 4,000 PackBots with US troops. She also ran iRobot's financing projects, which included raising $35 million in venture capital and a $75 million initial public offering. Greiner holds a bachelor's degree in mechanical engineering and a master's degree in computer science, both from MIT. She was presented with an honorary degree by Worcester Polytechnic Institute in 2009.

Greiner is highly decorated for her visionary contributions in technology innovation and business leadership. She was named by the Kennedy School at Harvard, in conjunction with US News & World Report, as one of America's Best Leaders. She was honored by the Association for Unmanned Vehicle Systems International (AUVSI) with the prestigious Pioneer Award. She has also been honored as a Technology Review magazine "Innovator for the Next Century." She has been awarded the DEMO God Award and DEMO Lifetime Achievement Award. She was named one of the Ernst and Young New England Entrepreneurs of the Year, invited to the World Economic Forum as a Global Leader of Tomorrow and Young Global Leader, and inducted in the Women in Technology International (WITI) Hall of Fame. Greiner is a trustee of the Massachusetts Institute of Technology (MIT) and the Museum of Science, Boston (MoS). She is on the Board of Visitors for the Army War College (AWC) and the Army Science Board (ASB). She has served as the elected president and board member of the Robotics Technology Consortium (RTC) and as trustee of the National Defense Industrial Association (NDIA).

Brett Stern: Could you give me your background? Where you were born, your education, and your field of study?

Helen Greiner: I was born in London, England. I moved to the States with my parents when I was five. I went to MIT, where I studied mechanical engineering but took many of the electrical engineering classes. In grad school, I studied electrical engineering and computer science, also at MIT.

Stern: And when you were growing up, were you inventing anything?

Greiner: When I was eleven, my dad brought home a TRS-80 computer—one of the early personal computers—and it was supposed to be for the family. But it soon became mine, because I was the only one who had patience to learn how to use it. I was making just the kind of things eleven-year-olds would make: video games, space invaders, keyboard races, etc.

Stern: Jumping forward, could you define the technology or your inventions in technical terms, and then define them in layperson's terms?

Greiner: I like to build integrated robot systems, or "cyber-physical systems," that are able to negotiate unstructured environments using dynamic sensing and onboard intelligence. Robotics encompasses many of the other disciplines, like artificial intelligence, dynamic sensing, and electrical engineering. You'd be hard-pressed to come up with an area that robotics doesn't use.

Stern: Could you define it in one sentence in very simple terms?

Greiner: I guess I'd have to say that I build practical robot systems.

Stern: When did you get the motivation for this field of study and what gave you this direction?

Greiner: I saw *Star Wars* on the big screen when I was eleven. It was 1977, and I fell in love with R2-D2. If you've seen *Star Wars*, you know that R2-D2 is more than a machine. He has a personality, he has an agenda, and he is one of the main characters. It was the first time that I saw potentially what a machine could be. Now, we're not there yet, but my motivation is R2-D2.

Stern: What was the state of the art, prior to you going into this field?

Greiner: When I was in school, there were a lot of robot labs and demos, but there weren't really any practical robots in people's hands. There were more what I would call "lab queens." Really cool stuff. And you can learn a lot, and discover a lot, and invent a lot, but there weren't a lot of practical robots that really did a job for people—that really did work.

Stern: When you were in school, were you doing professional or commercial work?

Greiner: I was an intern at NASA's Jet Propulsion Laboratory, so I was working on robots for space applications.

Stern: At what point did you decide to go the private-funded route as opposed to a corporate or government setting?

Greiner: When you looked at what was happening in robotics, I felt like what was going on in the laboratories wasn't really going anywhere. It just stopped when the grad student left or when sponsorship funding dried up. The researchers would work on it for five years, get their PhD or publish a paper, and then, "Oh, it's done." I wanted to do something that was more permanent.

Stern: Was the work collaborative or were you working alone?

Greiner: It's entirely collaborative. There is value in collecting ideas from a lot of different sources and combining them. Having a diverse team with different ideas and being able to pick the best ones is critical for successful design.

Stern: What do you think you bring to the project, then?

Greiner: I've got a very practical bent, so when people are proposing many degrees of freedom or very complex architectures, I'm able to say, "Hey, there's a whole downstream issue with those ideas." My philosophy is what they call KIS: Keep It Simple. That's what has allowed practical robots to get out into the world.

Stern: When you're starting a project, do you see the solution or do you see the problem?

Greiner: I think I see the problem. I might be a little disingenuous about that because everything has got a robotic solution for me, not a different solution. But really looking at the customer's needs first, instead of the robotic technology, has distinguished iRobot Corporation. For example, folks who buy a Roomba vacuuming robot—they don't care that it's a robot. They don't care what kind of sensors it has. They don't care what kind of microprocessor it has or how cool it is. They only care about one thing, and that is does it vacuum well? Is the floor clean when I get home?

Stern: You've done a lot of work in the military sector. Do they care about the guts of a robot?

Greiner: The military research sponsors do, but the guys in the field who are using it, not really. What they care about is that it starts up quickly, that it works each and every time they turn it on, that it doesn't require a lot of maintenance, and that it gets the job done. It succeeds in the mission. Other than that, they are busy, they are being shot at, or they are in a dangerous situation. They just need it to work.

Stern: When you're starting a project, do you have much input from these people, whether it's the military side or the consumer?

Greiner: Absolutely. I think that's essential, but you have to kind of work both ways when you're doing something completely new. It's not like the next generation of a product like a cell phone. You can't just say, "Well, we're going to do this, only faster or smaller."

A good example is when we first ran focus groups for the Roomba. We didn't show the focus group members our prototype at first. Instead, we just asked them if they would like a robotic vacuum and the answer was a resounding "no." They imagined the Terminator pushing an upright vacuum around. It wasn't until we showed them the initial design concept that they said things like, "yes, I would buy that," or "it can fit in my closet," or "it is so small that it won't damage my house," or "it is not scary like I was imagining." With a completely new product, you have to listen to the consumer, but you also have to show them new ideas to evaluate. They didn't like the idea of Roomba before we answered these questions: how much does it cost, what is it going to do for me, and where do I store it?

Stern: Do you think such feedback changes the design direction? In one sense, the consumer is going to give you money directly, compared to a soldier who is just consuming the product and doesn't really care about the financial side.

Greiner: No, I don't agree with that. Soldiers are responsible for the equipment they use [they sign for it and are responsible if it does not get returned], and they get can get attached to it. We had one Marine who returned to the depot holding the robot. It was in pieces in a box, and he asked, "Can you fix it?" The unfortunate answer [in such cases] is "no," if it's been blown up.

It turns out he had named this particular robot. He called it Scooby-Doo, and it had done fifteen IED [improvised explosive device] missions. He credited it with saving him and his buddies, so he grew really attached to it. I think soldiers, just like consumers buying a Roomba, buy it as an appliance. But when they get it home, [they see] it has some of the characteristics of a pet, so they start naming the robots as well.

Stern: Do you think it's important or a human necessity to somehow personalize it with a name?

Greiner: I don't think it's a necessity—it's just what happens because, again, it has the characteristics that people are used to in living things. Now I'm not saying the robots are living, of course. But some of the characteristics—moving around the house, responding to the environment in real time, having kind of a personality—are those that people associate with living things.

Stern: Do you think humans need some type of feedback from the robot?

Greiner: I don't know that they need it—but they like it. It helps them understand it's doing a job for them.

Stern: Do you design or engineer any type of personality into the object?

Greiner: Oh, yes. But, it's function first. Concentrating on cute instead of capable is backward. The product really has to do the job well because customers are buying it as an appliance. After designing a highly capable product, then we get some leeway to help to promote that bonding with the device. You would be surprised at how many quarrels we got at iRobot when customers called to say, "Hey, my robot's broken. Can you fix it?" And we reply, "Yes. Send it back," and they respond, "No, no, no. Not Rosie. You're going to come over in an ambulance to fix it, right? No, I don't want another one. I want Rosie fixed."

Stern: So in a sense, they want a doctor making a house call.

Greiner: Yes. They want their robot fixed rather than to turn it in like a cell phone, get a new one, and move on. So there is something unique going on with the robot technology.

Stern: I think over the years, people become very attached to their cars.

Greiner: Yes, I think so. And some people do name them. I think that if you told most people, "I'll give you a new car," they would turn it in.

Stern: That's true.

Greiner: They'd probably take the deal.

Stern: Could you explain your ideation process? How you approach a problem and how you get to the solution?

Greiner: It is matching need with the price of technology. We can build many cool and capable robots, but they will not be purchased because they use technology that is too expensive. So we start with the need and system architecture. We evaluate the probable cost of the system. We see what we need and what we can leave out. We see which expensive components can be redesigned to use cheaper technology.

If we get to a complete system architecture that works, we start performing more detailed design. If not, it might be best to change the constraints or try a different market need. Being able to pick areas ripe for sales today as opposed to twenty years from now is a tough but valuable skill.

Stern: Do you do you consider yourself an inventor first?

Greiner: At the beginning of iRobot, I used to go to the lab and build all the robots. At that time, I would have considered myself an inventor. Now I consider myself an entrepreneurial inventor who has to put together the team and make sure that they are attacking the right problem, understand the customer need, feel empowered to be creative, are heading in a practical direction, resist feature creep, and designing a unique product that meets a market need.

Stern: Do you think, wearing that inventor's hat, you find that you need any particular skill sets that are different than being an engineer or a scientist?

Greiner: I think you have to be able to make that jump from something that doesn't exist but could exist. A lot of folks have trouble with that. They give you all these reasons why it hasn't been done before, rather than jump in and imagine the world the way you want it to be.

Stern: Do you feel most people are afraid to take that responsibility?

Greiner: No, I don't think it's fear of responsibility. It is fear of heading down the wrong path. You know, you can always come up with reasons not to do something. It's hard to make that jump. Being able to say, "what's going on is good, but we can do better" is key.

We really have to work on letting kids do this. In school, you are taught to analyze. You are not inventing. You are given problem sets and you solve them. And that's what engineers can do, but inventors come up with something completely new and the first step of that is being able to say, "Well, why not?" Then they think through the reasons why it hasn't been done before. Good inventors come up with ways to address these challenges rather than throwing up their hands and quitting.

Stern: It seems what you are saying is that the inventor is not afraid to fail or is almost looking for a failure.

Greiner: Well, I wouldn't say looking for a failure, but failing is part of inventing. As Edison said, "We now know a thousand ways not to build a lightbulb."

Inventors in general believe they are going to succeed. We have to really believe to push through some of those speed bumps along the way and make sacrifices to pull it off.

Stern: Do you think that beliefs are in one's self? Or that the beliefs are in the technology?

Greiner: You really have to believe in yourself. Technologies come and go, and there can be different innovations at different times. The belief is that you can make the right decisions and really pull this off. You might change from one technology to another while you are inventing, and that shouldn't really matter. It is really the new capability that you are bringing to the world that matters.

Stern: Would you like to tell me any of your biggest failures?

Greiner: Yes. My biggest failure. I am calling it a failure, but I think I was just ahead of my time. We tried to put an Internet-connected robot on the market in about the year 2000. This robot was would allow a person to visit a certain location over the Internet—it was mobile with eyes and ears. We had a pro-totype and did great demos.

It could have worked except we were much too early—the speed of the Inter-net and the adoption of home networks were not yet to a point to support this technology. But it will come. Several companies now have such products on the market for commercial uses.

Stern: You talked about innovation. Do you have any particular definition of that word?

Greiner: Imagining, then creating, something that did not exist before.

Stern: When you are doing the inventing part, do you have any particular tools that you use? Do you sketch? Do you prototype in 3D? Do you have any favorite materials or technologies that you like to use for the process up front?

Greiner: I probably shouldn't admit it, but I use whiteboards and spreadsheets. Whiteboard for sketches to have a group come up with new ideas and feed off each other's creativity. Because the invention is more than the concept. It is asking how the customer is going to use it. How much does it cost? How much value does it provide to them? All of that has to be wrapped up in some format—and Excel is a nice little tool for coming up with costing and then deciding what technology fits the cost profile. So, I don't mean to get all business-focused on you, but inventions only see the light of day if they are purchased by people.

Stern: No, that is okay. I want to hear your process.

Greiner: Invention without the sense of who you are going to sell it to and how much are they going to pay is not as interesting to me. Probably because of all the really cool robots that exist in the labs that never even come close to commercial adoption.

Stern: So you are looking at the commercialization side right up front.

Greiner: Absolutely. I think if more robotics inventors did this, we would have more robots out there in the world.

Stern: Do you work on a computer or are you working with a pen and paper?

Greiner: Oh, it is all on computers. Robots are highly complex, integrated electromechanical computer science, sensory, systems. There is not just one person inventing them. There are teams of people working together. So there has to be design packages that everyone can access and share ideas through.

Stern: Do you have any methods of dealing with these groups of people in getting everyone to interact? Do you have any methods for motivating them to come together with their best work?

Greiner: I think people do their best work when they are involved at the beginning of the process. If you get folks discussing the need from the beginning and coming up with the solution, they become the core team. That can really help with motivation. It is not as if someone tells you, "Do this piece of the project." It is seeing the whole concept, the reasons why a particular route was chosen—and this is my piece of it.

Stern: How would you say that the technology that you have worked on so far and the products you have put out there—how have they influenced society?

Greiner: There are a lot of examples from delivering to consumers and the military. At iRobot, I get letters from Roomba users all the time saying, "This is great—it allows me to save time." Or even better, "I don't have to drag my larger vacuum out of the closet. I can still do the vacuuming without having to call on someone to do it for me, and that allows me to maintain my independence, which is important to me."

Before our military robots were deployed, standard operating procedure was to send an EOD [Explosive Ordnance Disposal] technician up to a bomb. Now it is standard operating procedure to send a robot. I will give you one story. I was at the Army War College giving a talk a few years ago. Afterward, many of the Army colonels in the audience came up to me to say thank you for the robots that were in the field saving lives. One colonel, who was with the 63rd Ordnance Battalion EOD in Iraq, told me my robot saved eleven guys on one mission! It doesn't get better than that.

Stern: You have produced products for the consumer and military sector. Do you have a preference?

Greiner: I think they are both great areas for robots, and there are so many more. I am proudest of the work that we have done for the military, because it has been successful in saving the lives of hundreds of soldiers, sailors, and Marines. So although Roomba is probably affecting more people, the larger impact is soldiers able to return home to their families instead of becoming IED casualties.

Stern: To what extent does financial reward or opportunity play in making decisions to go into consumer vs. military projects?

Greiner: You definitely need to consider the business case before you start. Fortunately, there is a great business case for robots in both the consumer and military spaces.

Stern: Could you talk a little about the intellectual property side? How much is driven for the demand to have patents or something proprietary?

Greiner: I think the most important thing is to get the invention to market first and gain the first-mover advantage. I look at some patents in our field and ask, "How can you patent that?" since it is well known in the field. I would be an advocate for changes in the current patent system to allow third-party review as part of the process [to increase the quality of the patents that do survive].

Stern: You have numerous patents, correct?

Greiner: The companies that I founded have numerous patents, but I don't have any personally.

Stern: In the work that you are doing in your career so far, do you have mentors? Where do they come from?

Greiner: I haven't had any mentors because I founded iRobot with two business partners right out of grad school. Basically, we had to mentor ourselves.

Stern: Do people come to you looking for you to be their mentor, since you have established this method for putting new ideas out there?

Greiner: I hope I do provide some encouragement and feedback to folks. Usually it goes along the line of keeping it simple. Find an idea that you can pull off without multiple new components and complexities, etc. Then build up complexity from a base of success.

Stern: Are there any inventors or inventions that you admire, whether the person or the product?

Greiner: I have two or three that come to mind. Mary Anderson, who invented windshield wipers in 1905. Stephanie Kwolek, who invented Kevlar. And Admiral Grace Hopper, who invented compilers.

Stern: Why the windshield wipers?

Greiner: Well, these inventors had to overcome more than the technical and adoption hurdles typical for any inventor. They also had to overcome gender stereotypes and biases that were even more prevalent in the past.

Stern: Would you say there is a reason why women don't get more involved in the inventing or technical world?

Greiner: Women have to overcome many stereotypes and biases to get into positions where they can invent. Even obtaining financial backing, for example.

Stern: Do you think that is changing now? Would you like to give any encouragement for women out there?

Greiner: Oh, yes. What I would tell younger folks is that being an inventor is a great way to change the world for the better. Many young women tell me that they want a career that has a positive impact on people. They want to do things that help people and the planet. The best impact that they will be able to have is to become an engineer, design a dam, make more energy-efficient buildings, come up with new ways to store energy, come up with a solution to the world's potable water problems. If you want to have an impact, well if you want to help people, become an engineer or scientist and then an inventor.

Stern: I am assuming that is not just limited to women. It is to men as well.

Greiner: Yes. But, I think that the message does tend to get to boys a little bit better already. For example, a Best Buy ad campaign in 2012 that highlights only male inventors. They kicked it off at the Super Bowl and have been running it during primetime TV.[1] So boys are already getting millions put into the message that they can be inventors.

---

[1] www.youtube.com/watch?v=VyCDzLebBpc

Stern: It is easy to give the example of Facebook or Google as the commercialization side, but the reality is that there are very few people that actually get to that level.

Greiner: They have taken a whole bunch of inventors with them. But, I would admit not everyone will get to that level. Regardless, if you study engineering or science, you have got a good base to go out and make a great living, support your family, do well in your community, etc. So it is really a no-lose situation. It is also a good foundation to go into another field like law, medicine, bioengineering, finance, investing, etc. And you could be one of those skilled, motivated, and lucky people who achieve stratospheric success as an inventor.

Stern: How has the success of your work changed your life? What opportunities do you feel have opened up for you?

Greiner: Because of prior success, I get to spend my time taking on new challenges. My motivation has always been to get robots in people's hands, and that has happened—a dream come true. There are six thousand military robots now and six million Roombas in people's hands. But, I want to keep on building. I like creating things the world hasn't seen before. That is why I am doing another start-up in robotics.

Stern: Well, can you talk about your next project?

Greiner: Not really, no. It is still in "stealth" mode. But I can say we are focused initially on UAVs [unmanned aerial vehicles]—flying robots!

Stern: Where do you feel that robotics technology has not been applied yet?

Greiner: Oh, there are lots of great places. Just off the top of my head, agriculture and transportation. We don't have automated farms or real driverless cars yet. There is a lot that can be done robotically in exploring the universe or exploring underwater that hasn't yet been taken on. There is a lot you could do with having networks for early warning systems for natural disasters in the oceans and the sky. Any time you have a task that is dull, dirty, or dangerous, I believe a robot should be invented to do the job.

Stern: Are there any fields that you think robots shouldn't be used?

Greiner: I think authoring books would be the one field. No, I am just joking. I don't really see any. I think that robots will cooperate with humans just like humans cooperate with computers today. I don't think of myself as less informed because I use a search engine. Using computers for what they are good at and using the human brain for what it is good at is a powerful combination. The same is true with robots. You use a robot for what it is good at in the shorter term, at least, and use the human brain for what it's good at.

Stern: What about the emotional side between people?

Greiner: There is great work around the world in these areas—robots that use effective interfaces. For example, understanding a user's expression. It is not my expertise. I believe in the simplest interface, whether it mimics human-to-human communications or not. Roomba coming out to clean everyday on its own is a more effective interface than if it needed human commands.

Stern: You have talked about the projects from a business standpoint. You went to school for engineering. Where did you get this business expertise?

Greiner: The school of hard knocks.

Stern: It's a good place to study.

Greiner: No, it's challenging. Because when we first started iRobot, we were always doing great robot projects, but most of them didn't go anywhere. We were always convinced that they would. It was like that for the first eight years. We didn't try and raise any capital. We also didn't have money in the bank to pay payroll at the end of the month for five years.

We did some great inventions and some wonderful demonstrations of really exciting technology. But what has been more rewarding to me is when folks have your robots in their homes. When a guy tells you, "this robot saved eleven soldiers on one mission"—that is why I like concentrating on areas where you have that kind of an impact, rather than what I would call "invention for inventing's sake."

Stern: Do you have any overall advice or anything specific that you would offer to an inventor?

Greiner: I would say invent in an area that you are passionate about. I think most inventors do. They see a need and they get passionate about it. But there will be roadblocks along the way, and in order to plow through them, you have to be passionate about the need you are addressing. If you are not, you will not get up every morning to try and solve all the roadblocks that come up along the way. When you are passionate about something like I am, along with my teams, anything that comes up—you will figure out a way to get to your final goal.

Stern: Outside of robotics, do you have any favorite products or tools that you like having around your house? Things that you find indispensable?

Greiner: Indispensable? Well, I go [everywhere] with this cell phone and a tablet.

Stern: How about things that are dispensable?

Greiner: Well, I would have thought my tablet would be that way until I started using it, and now it is indispensable.

I will try to think of one off the top of my head—projection TV with good sound, or automated blinds and lighting. I would say they are probably luxuries.

Stern: Anything nontechnical that you really like having around?

Greiner: I like flowers and plants.

Stern: Any particular type of flowers?

Greiner: Well, I mostly like roses, but I keep them outside growing and not indoors in vases. I had a gardening phase, but I am not doing that so much right now because of the start-up company.

Stern: Do you have any sort of big view of what the future will look like in your field?

Greiner: I would say that physical labor as an option, not a requirement of life. You might like doing certain things, like gardening, for example. More power to you, but it's better as a choice not a chore. With Roomba, vacuuming the floor today *is* an option.

Stern: What do you personally do for fun or distraction?

Greiner: I snowboard. I kiteboard. I wakeboard. I kayak. I recently returned from a hiking trip on Inca trails through Machu Picchu.

Stern: Did you get to the top? Is there a top?

Greiner: I got to multiple tops. Machu Picchu is actually lower than the highest. But I also did Kilimanjaro three years ago, and I did get to the top.

Stern: When you are hiking out there on the trail, does it give you a chance to think about other things besides hiking the trail? The projects you are working on?

Greiner: Machu Picchu was very interesting because there is a lot of engineering involved in building the trail and sanctuary. I think it's an inspiring example of human ingenuity. I have been trying the whole time to figure out—and I am not an expert in South American history—but it is claimed that the Inca Indians [Quechuas] didn't have a written language or the wheel. I am just trying to wrap my head around why didn't they build wheels and why they didn't write things down, especially since there are other cultures in their areas that did. So I guess I was mostly thinking about their culture and trying to get away from the robots for a week.

Stern: Do you find it necessary every once in a while to get away from the subject matter?

Greiner: I don't do it often, but I am trying to take on one big thing, mostly because I am not sure if my body will withstand doing it ten years from now. I would hate to not have the opportunity to go and climb Kilimanjaro, see Machu Picchu, do the Inca trail. So I am trying to make sure that I make time because my natural tendency is to be a workaholic.

Stern: What does your daily routine look like?

Greiner: I get up and go to work at the company. There is so much stuff going on, and one of the things about a start-up company is that you are doing something different, so it is not standard routine. It is whatever is going on that day. Whatever you have to push through. Whatever went wrong. Whatever is coming up. And solving problems before they exist. So building a company is a lot like inventing a robot.

Stern: At the end of this, do you ever plan to retire?

Greiner: I would say yes, but I don't think I would plan to do nothing. There are so many other things to do. Learning something new every day is important to me. Giving back, encouraging little girls who are bombarded with media messages—like shoes are the most important thing to women—to take on big challenges.

Stern: Any final words that you want to actually get out there to people who are aspiring inventors?

Greiner: I think I have already said it, but getting to a field that you are really passionate about—whether it is robotics, or bioengineering, or civil engineering—in the beginning you will run into roadblocks. Every project does. Some of them might be substantial and daunting. If it were easy, someone else would have done it. It is never going to be easy, but if you are passionate about it and it is what you want to do with your life, you will figure out ways around the obstacles, and you will be motivated to get up every day to work on it. My job has never seemed like a job because my goal was to build robots, and I am very fortunate that I am able to do that every working day of my life.

Glen Merfeld

Platform Leader
Energy Storage Technologies, General Electric

Glen Merfeld is the platform leader for Energy Storage Technologies at General Electric. The span of his position includes all forms of industrial-scale energy storage. He is responsible for program activities and strategic developments across GE Global Research, and for integration with other GE businesses. Merfeld also assesses emerging technologies in support of GE Capital Investments. His platform activities are closely tied with the energy storage business, including continued sodium metal halide battery developments and their translation to commercialization.

Previously, Merfeld was manager of the Chemical Energy Systems Laboratory at GE Global Research. His group led the advancement of new chemistry-based energy storage and conversion technologies. Their efforts included electrochemistry developments for novel battery technologies to enable the electrification of transportation, and improved stationary power quality in potential applications ranging from grid utility, to renewables, to uninterrupted power supplies. This research group led the design and synthesis of molecules that make organic light-emitting diodes (LEDs) brighter and more efficient. The group also created material systems to forestall the environmental degradation of thin-film solar devices and packaging systems.

Merfeld and his team have played a foundational role in advancing the sodium metal halide battery technology that is now the basis for GE's new battery business, GE Energy Storage Technologies. It is currently ramping up a manufacturing facility in Schenectady, New York. Merfeld's team was recognized with the 2010 Whitney Award—GE's highest recognition for breakthrough technology innovations. Beyond technology development, Merfeld has played a central role in shaping the battery business opportunity for GE through market assessment, acquisition strategy, and technology evaluation.

Merfeld joined GE Global Research in 1998 after earning a bachelor's degree in chemical engineering from Northwestern University and a PhD in chemical engineering from the University of Texas at Austin.

Brett Stern: Could give me your background? Where were you born? What's your education and field of study?

Glen Merfeld: I was born in Iowa. I grew up on a grain farm in northeastern Iowa. I received an undergraduate in chemical engineering at Northwestern in Chicago, and then I went on to Austin, Texas, where I received a PhD, also in chemical engineering. About fourteen years ago, I joined GE. I came to upstate New York to the research center here and did a lot of work early in my career on plastics, on new polymer developments. My background is polymer physics and material science, and so I did a lot of work on developing products that got launched into the composite and the coating area. Then I worked on increasing materials for energy conversion and energy storage. This was maybe seven, eight years ago.

Then I started managing The Energy Storage and Conversion Laboratory here at the research center. We've done things like developing packaging materials for solar cells to make them more environmentally robust so that you understand how they degrade in the environment. We've worked on organic molecules that you use in organic light-emitting diodes. My lab also did a lot of the foundational work for advancing a battery technology called sodium metal halide.

Stern: Growing up on a farm, would you say you were inventing things as a kid?

Merfeld: I think so. That was kind of the environment that we had to work within, because farming is a rather creative endeavor. The farmers that I was always exposed to were the type of people who were willing and eager to take risks. It's not an occupation that's without those sorts of hazards. So maybe it was sort of that environment, but certainly, it was a very hands-on way to grow up. We had a shop right there on the farm where we welded and built and destroyed things, so I feel really fortunate. There is certainly discipline that is needed to be a farmer. I think to be a good inventor—maybe this ties back to Edison—I think you have to have a lot of discipline and you have to be very persistent. I think there are a lot of examples here at the research center for GE where we rely very strongly on that approach.

Stern: Could you talk about the work/inventions you're doing now?

Merfeld: Specifically the battery. Energy storage today has gotten a lot of attention for good reason. And that's because energy storage is a powerful way to make better use of energy resources. To put this in simple terms, it's analogous to how you might drive your car. I think we all appreciate you get your best fuel economy, the best use of your fuel, when you're running steady state, when you're at cruise on the highway, right? When you're not hitting the brake and you're not slamming on the accelerator, right? That analogy is perfectly inflatable when you start inserting energy storage into the mix. So if you can take some of those excursions, either braking or accelerating a car, and you capture them with a battery, it allows you to make better use of your energy resources.

And that whole analogy is completely parallel to when you start thinking about stationary electricity, the stuff that you and I consume as soon as you flip on the switch. Instantaneously, resources somewhere have to come online to generate that electricity. We're all familiar with the fact that there's very much a diurnal-type cycle associated with human consumption of electricity, and that causes us to bring on and take off generation assets. Sometimes those assets are not always the most efficient. The ones that are the most responsive are not always the most efficient. What's exciting about energy storage is that you can do it efficiently and effectively, and safely and reliably. You can do some pretty amazing things to help manage our energy resources.

Stern: How did you go from energy storage systems to applying it to locomotive trains?

Merfeld: That actually goes back to how we got started here. What I think is key to pulling off an invention is having a clear target. It may not be ultimately the final target, but I think it's important for inventors to stay focused, because if your goals are constantly shifting, you end up spinning your wheels. So early on, we were not setting out to develop a whole new battery and ultimately a new business for GE—which is what happened. What we were challenged with doing about eight years ago was solving an issue that our transportation business had with locomotives on the rail today.

These are electric-propulsion vehicles. All the drive is done with electric motors. There is a diesel generator that creates electricity. The operation is when you go to brake one of these large vehicles that have thousands of tons of mass behind them, you're taking a wall of kinetic energy, and you're simply turning it into heat. There's a large grid resistor on the top of the engine and you dissipate all that kinetic energy of heat. There is an opportunity to could capture some of that, very analogous to a Toyota Prius. If you capture some of that braking energy, then you can use it during acceleration or cruise, and that allows you to save fuel. It allows you to reduce emissions. That's where we got started. We were targeting ways to create a better energy storage device that would address that challenge. We run locomotives in very cold climates like Alaska, or very sweltering climates like South America or Australia.

Stern: When you started this project, did you start with written problem statements? Do you sit around and brainstorm where to focus the potential activities?

Merfeld: Yes. Our process is very deliberate. To a large degree, we want to have a good understanding of what the problem is that we are trying to solve, because scientists have great ideas and there is never a shortage of that. I think the key way that we approach research, certainly on more applied programs, is that we want to get aligned with the right problems. I think it's fair to say that a lot of times the biggest challenge in doing innovation is tackling or asking the right question. Identifying the right problem. So we spend a lot of time doing it.

I'm pretty confident whenever we can corner the real problem and the real challenge that we're trying to address, that's almost the easy part. It's trying to pin down what you're trying to tackle. So we spend a lot of time doing that—and it's a tremendous amount of work.

Another part of my role is that I look at emerging technologies. Our company makes investment on emerging technologies, so often I see companies, small start-ups, doing tremendous technology, but it's completely misaligned with the opportunities that are out there. So that's why I think our approach might be distinctive.

Stern: You are a manager now. How hands-on do you get with the actual scientific work or engineering work?

Merfeld: There is a whole spectrum of contributions that I think we need when we do innovations. I'm not the guy that's in the lab doing the experiments anymore. I spend a lot of time with the team, looking at the results and trying to understand what it means. A lot of my role now is even a further step removed, by making sure we have the right strategy. Making sure that we are anticipating where we need to be in a year, or two, or three to put together a total solution.

So a lot of times, my role is challenging our bench scientists to think about the next problem, and how to make our capabilities even more sophisticated in predicting where we should be. It's been tremendous to be part of the battery story, for example. Because I can share with you that six years ago we started to do things like electrochemical modeling. At that time, it was interesting, but it wasn't really advancing our capabilities the way that six years later they are now.

That's an example of where we had to anticipate that we needed to get to the point where we have models that are predictive, where we can change the design of our cell or alter the chemistry and predict how that's going to manifest—how it might change the dynamics of the industry. We have system-level models that allow us to anticipate the value that we bring to our customer if we make a change to the design of a fuel cell.

Stern: You talk about the customer. Do you have input from the customer? And who is the final customer?

Merfeld: So in locomotive applications, or in industrial transportation–type applications, those customers can be the folks who actually buy the locomotive engines, haul freight on the track. We're fortunate in that we have a business that's been on the ground for probably close to eighty years now. So our customers are always the folks at the end that pay the bill and that use the asset. Now we also have probably as demanding of a customer internally here, our counterparts that are part of the various GE businesses. These are the engineers, and the deep scientists, and the marketing leaders that help keep us focused on answering those challenges.

Stern: Does the customer come to you with the problem, or do you go to the customer with the potential problem? With this battery project, how did that come about?

Merfeld: So in that case, the customer, our transportation business, was working side by side with the research center. We both recognized the need and the opportunity there, so it was very much a joint venture, trying to understand the best solution to hybridizing a locomotive. We built an energy storage business, and the real launch technologies are applications different from those that we initially focused on. It was fantastic for us to have such a driven target for a hybrid locomotive, but as we got smarter about the marketplace and we got smarter about the technology capabilities, we got to the point where we realized that not only to make the technology sing, but also to make the economics work, we needed to get volume. The amount of locomotives that we were going to sell wasn't going to justify putting a large investment into a factory.

It was somewhat fortuitous but I think it's also reflective of the versatility that I see in some of our leadership style: we recognize that if we have a good technology that can outperform other solutions out there, we should think about starting a whole new business where we can take this battery and use it to address stationary applications. So a lot of the launch of the energy storage business now isn't on the hybrid locomotive, but rather it's focused on emerging markets where we can sell volumes today. These are things like telecommunications, things like grid-type applications, things like critical power support, and applications like hospitals and data centers.

Stern: So the locomotive project gives you the opportunity to learn and then eventually scale up to other applications?

Merfeld: Absolutely. I think in the back of our minds, at least from a leadership and management standpoint, as long as we picked a tough target initially, to scale it back for a nontransportation application, was an easier thing to do than vice versa. In some ways, you want to make sure you put that challenge very far out there. I've come to accept that our scientists are incredibly capable of tackling almost any challenge. I think the bigger challenge sometimes is making sure it's hard enough, and making sure that we're swinging hard enough for the real big opportunity.

Stern: In the corporate GE setting, do you have access to all the engineers and scientists? Do you ever sit down together with all the different heads of different departments and say, "I could use this, I could use that?"

Merfeld: Absolutely. That's just part of how we structure our organization. As we are going through the process of commercializing the technology, there are skill sets that we need at the table that we'll go out to our aviation business, for example, or we'll go to our energy business, and we'll just pull in the people who have decades and careers of experience to bring to the table. It's just incredibly powerful that we have that ability as a broader company. And then just here at our global research centers, that's the whole spirit of how we staff our organization and our projects specifically.

You know, when I think about tackling a problem, I almost deliberately want to pull in people who aren't biased by a preconception of how to solve it, and we certainly have that right here on our campus, and across the globe—people who have skill sets that are incredibly diverse. We have mathematicians, physicists, nuclear scientists, chemical engineers, chemists, and you name it. If you can find it on a college campus, I guarantee you we have not only that represented here, but we probably have somebody who's been doing hands-on innovation in that specific expertise for twenty to thirty years. And it's just incredibly powerful, because we come up against problems that I think, to a large part, a lot of people would look at and just walk away. I'm fortunate because I can call up my colleagues here and very quickly pinpoint who's the best person in the company to bring to the table. Within twenty-four hours, they are at the table and we're knocking down problems.

Stern: Obviously, there's a commercialization side that you're going after. Do you start thinking about that right from the beginning, and do you have goals along the way to reach for the commercialization of a project?

Merfeld: I think the answer is absolutely yes. And I'll share with you an example that really exemplifies that. In this battery program, back even six years ago, right here at our research center, our team did a spectrum of activities on advancing that technology. It went all the way from the fundamental electric chemistry in a cell to the power electronics and packaging of the battery, all the way through, up to manufacturing.

Right here at our research center, we have a whole organization that's about manufacturing technologies. Six years ago, long before we ever made the decision to pull the trigger on building a factory, those experts were at the table. They were at the table anticipating supply chain challenges, thinking about manufacturability, thinking about resource management, sustainability, and even six years ago, our material scientists and chemists were taking feedback and working hand in hand with those folks. We even did a lot of simulation of what a factory would look like, what the tolerances of the different pieces and components were, and again, that all took place long before anybody ever made a decision that we were going to put a factory on the ground.

Stern: Obviously a long-term project. How do you keep yourself and everyone else motivated along the way?

Merfeld: I think, foremost, the people here that work on these projects are very driven to learn and to advance science. So it's, quite honestly, an easy challenge to keep people motivated. They instinctively and internally are passionate about innovation. When you're slogging out on a tough problem and a program that has had legs and been around for more than eight years now, you're constantly trying to make sure that you're challenging yourself. I think—and that's where I've found I play a big role—I see when people have gotten a little comfortable with their space, and so I challenge them to work in a different space. We constantly restructure our teams. I sit down with the project leaders, formally sometimes, but I am constantly pushing them informally, to question whether we even have the right people. Is it time to change up our batting order? Should we bring in somebody new?

We had an example at the end of last year where we had come up against a very challenging defect mode and we pulled together some of the physicists from the electronics part of our organization who weren't involved. We also pulled in folks from different parts of the GE business, and in four hours, we had six new ways to think about how to tackle this issue. One of those solutions was actually dead-on, and a week later, we had a whole new solution. So keeping people excited by making sure they're working on relevant things is part of it. But boy, if you give them a big challenge and you give them the resources and a fertile ground, it's just incredible to be part of that.

Stern: Do you have any particular definition of innovation?

Merfeld: I'm an engineer. The way I think about innovation might be a little bit different than how a scientist [would think about it]. I'd like to think I'm a scientist, but when you're in an organization like we have here, there is a group of those who are more engineering-minded, and a group of folks that are very scientist-minded. And innovation to those two groups, I think is very different.

As an engineer, it's not so much about creating a new material that nobody's ever thought of before. As an engineer, it's focused on solving a problem. I remember when I was interviewing here at the research center, the common question—and I still ask this question today to people I interview—was what's so innovative about your thesis? What did you do that people haven't done before? Engineers, I think, are very comfortable with taking other people's ideas and bringing them together to develop solutions. That really motivates them. That motivates me. I want to start a new business and solve problems that actually are important to the world. Scientists are just passionate about solving that core underlying fundamental question. And when you team up that mind of a pure scientist with an engineer, I think that's an incredible strength.

Stern: Since you are working with both engineers and scientists, is there a different skill set or complementary skill set that those people need to be inventive? Is there a commonality among the people that you are working with?

Merfeld: I absolutely think those two different approaches are necessary. I think an engineer on his own won't get you where you need to be. Because you will come across challenges that require you to fundamentally get deep and really understand why something works mechanistically or why it doesn't work mechanistically. When you do that, you can open a whole new opportunity.

And I think the other is true, as well. I think there are times in research, historically, where it's been less applied and more perhaps academic, where you can do a lot of great fundamental work. But unless you create that connection between the fundamental work and that pragmatic engineer who can extract the value and turn it into something that somebody will pay for, these great ideas will go unrecognized.

Stern: Can you talk about the work you do? What tools do you use to ideate, to brainstorm? Do you sketch? Do you prototype in 3D? What are the favorite materials and technologies that you use to look at a direction for solutions?

Merfeld: We use anything that you can imagine. We do world-class modeling that's probably unmatched, where you can really understand and anticipate all the physics of ion transport and electron flux, and then get 3D visualizations. Many times the models may not be predictive in their own right, but they become scaffolding on which to test your ideas. And that's incredibly important.

One of my early mentors was a physicist, and that's what he taught me early on. He said, "You know, modeling, if you go at it and really expect it to be predictive, you're going to be disappointed. But if you use it as a tool to get your head around the visual picture of this problem you're trying to tackle, it's powerful." If you don't get so wedded to your concepts and your perceptions about what the problem looks like, and you have versatility and the ability to change that construct in your mind, you can respond to the data that you get. I think that's one of the mechanisms that we use.

We certainly do a lot of prototyping. Nothing better than getting your hands dirty in the lab. We try to create full-blown prototypes of the systems that we are developing. We do subscale systems that allow us to isolate subproblems. We take technologies that were born out of our medical imaging business, for example. You're familiar with computer topography CT: you want to look inside the human brain to see if the patient has a brain tumor, or tore the ACL in the knee. You can do that without cutting open the human body, fortunately. We take that same sort of approach and we apply it to batteries. So we actually can take CT technology and look inside our cells without cutting them open, and that's incredibly powerful. Because just like the human body, a lot of times in the process of tearing something apart to do a postmortem, you've destroyed the evidence and you've taken the problem out of its true environment, and so you've corrupted it.

We do things where we want to visualize and volumetrically render what's going on inside of our battery cells, and we can do that here. Anything's possible. We've taken our battery down to Brookhaven National Laboratory just two-and-a-half hours south of here on Long Island. We take our cells down there to take the high-energy X-rays that come off the synchrotron, and we'll do defraction to our cells. And what you see when you do this is interference patterns that are almost like a DNA signal, but the chemistry and those fingerprints allow us to understand spatially the time-resolved changes that happen in the chemistry. It's unmatched capability. Once you have these tools to look at the problem differently—and sometimes it's just somebody's mind that thinks about problems differently—that's when you unravel things and see things in a way that people haven't before.

This chemistry that we're commercializing, we didn't invent. It was first identified thirty years ago in South Africa. So we take no shame in the fact that what we've done is take that technology and advance it to the point that it's very reliable, it's very ruggedized, it has good energy density, and it is improved over what's conventionally out there. In that sort of engineering mindset, we're okay with the fact that we didn't invent that first technology. But we invented a lot of ways to make it better and actually solved commercially relevant problems.

Stern: In doing the research and exploring these different venues and having all these tools, do you ever get to the point where there's a failure? Can you talk about the failure process and what you do to get up and go after that?

Merfeld: Yeah. Failure—if you're doing the job right, if you're pushing yourself hard enough and pushing the technology—will occur almost every day. We expect it. Actually what's cool about failure is if you embrace it as an opportunity to dig in and understand, it really is an opportunity. We do that deliberately in the way we do research. We take our batteries, for example, and we stress them to the point of failure. We deliberately abuse them. We fail them, and then we tear them apart and we try to figure out where they were weakest. We attack that weakness and we try and understand why they are weak there and that's where you make them stronger. It's just how we do the research.

It's almost going back to Edison. Edisonian. You beat it up and are never satisfied with something, because everything has a vulnerability. Even when you've got that product and you've launched it, you want to be thinking about what's next in the pipeline. How can you make it better? The only way you can do that, sometimes, is by failing. Now throughout the last eight years in our program, there have been what appeared to be insurmountable challenges. You come up against problems where things aren't behaving the way they used to behave. You start scaling processes and you're not getting the same thing that you got when you were in the laboratory. These are big challenges that we often face. And I've got to tell you, six years ago those things *really* caused me to lose a lot of sleep. Six years later, I don't worry as much anymore, because I've just seen our teams, time after time, knock these things down. So I think it's just almost a belief at this point that nothing's gonna stop us.

Stern: So it sounds like part of this is a maturity, having enough experience to understand that there's a way to go forward.

Merfeld: Absolutely.

Stern: Before you mentioned Edisonian. General Electric has a heritage from Thomas Edison. Is there a ghost of Thomas Edison in the hallways that is watching over what you're doing?

Merfeld: I don't know about a ghost, but there's a great big mural of Thomas Edison standing in his laboratory in one of our galleries here, with the quote: "I find out what people need and I proceed to invent." I'm reminded of it every time I bring in a guest or a customer. We bring them into our front lobby and we have Edison's desk sitting right there. We have a display case that includes some of the medals that were won by the esteemed scientists that have preceded us here, two Nobel prizes included. Every time, without fail, whether it's a customer or whether it's a work partner, it's just this great respect and awe that people take when they get to touch the desk and then they see the picture of Edison on the wall. I think we probably take that for granted here, when you're so immersed in this environment. It's hard to imagine not having that sort of pedigree and genealogy that we can tap into.

Stern: In your career, as you have been growing and expanding, have you had mentors? How do you find mentors and what attributes do you seek in them?

Merfeld: Fortunately, I'm surrounded by folks that I consider mentors. I shared one with you previously. Early on in my career, I met a physicist who was just next door to me and had spent twenty years both at the research center, as well as in GE's plastics business. So there are mentors up and down the hallway here, and you have mentors that are very deep technically, who will be that touchstone that remind you constantly that it's really about the science and the engineering. Don't ever forget that. It's about the rigor and the discipline that it takes to do that well.

Around here, it's almost a peer environment where you constantly can't afford to be the person in the room who doesn't know the latest and greatest. People take pride in the fact of reading the latest science articles, or knowing what's going on in the world. I think that goes not only for science. For me personally, I spend a lot of time trying to understand what's going on in the business world, because I realize technology, great technology in its own right, is insufficient to be commercially successful. So I think mentors that I seek out now are not only the people that are technically very deep, but also people that have a good perspective on what it takes to be successful in the business world.

I've been fortunate that I had a neighbor that was the president of a local college, who gave me terrific career advice early on and I still stay tied into. And we have people here partially because of sort of the nexus nature of how we run research. We're constantly bringing in our business leaders. Our CEOs are often in here, and our exposure to the forces that they use to make their decisions are something that we share. And they are very open, and those types of folks are I think mentors not only to me personally, when you can sit down and spend some time with them one-on-one, but also I think that in a collective way it's just sort of how we try to structure research here in our interactions with our businesses. Whenever we bring a CEO in for technical reviews, they are always given an open seminar with the scientists—and we fill up a big auditorium. They tell us what they're thinking and what they're seeing, and then they open it up for questions. Some of the questions that you get are what you would expect: about how is the business doing. But then you can get very deep into the factors that are influencing the economy today and how technology interfaces with that.

Stern: It sounds like the commercialization is a big part of the process.

Merfeld: Yes, it has to be.

Stern: As the manager of this, you are dealing with the scientific side, but do you get involved in the marketing side or the selling side?

Merfeld: Absolutely. You know, a big part of my role, formally, is platform leader for Energy Storage Technologies. So that title is rather deliberate. I not only work very closely for this energy storage business, which is commercializing what we call the Durathon battery. But my role goes beyond that. My role is to help all of the GE businesses understand energy storage solutions. Some of those GE businesses are the GE Capital Investment Groups. So I spend a lot of time doing technical diligence on emerging technologies, grading that technology, developing tools that allow us to "rack and stack" technologies. If we can use these tools to make informed investment decisions, we can also use that same tool if one of our businesses comes and says, "Hey, I need a way to store energy in this type of environment with this type of attribute."

Stern: In a sense, it's an internal technology transfer that you're doing.

Merfeld: Absolutely. And that's a big component of it. There's a role that I feel very fortunate to play, as sort of an unbiased arbiter of what is the right solution. And that maybe ties in more to my engineering trade. The last thing I would want to do is take a technology and shoehorn it into an application because we could make money. That's insufficient. So that means sometimes you need to anticipate and understand if there are alternative ways to get to the same point that are more efficient or more effective or cost effective—that's the solution that we should seek.

Stern: Do you think there are fields of use that haven't applied the technology yet? What shape do you see the future taking?

Merfeld: I think the whole stationary grid space is exciting, underdeveloped, and in the early percolation stages of evolving. There's a lot of excitement about this space, and I think it could be an opportunity like the hybridization of transportation to get better fuel economy. Applying that same concept around how we do power generation for electricity is going to be pretty exciting. Whether it's distributed generation, whether it's allowing the increased penetration of renewables, whether it changes in the way we actually think, whether it changes the models that we can use to envision how you make and sell electricity, I think energy storage has the promise to change that whole space.

Stern: This technology is fairly macro. Do you see it getting into a micro level? I mean are there products that you will eventually make that the individual will use?

Merfeld: People have proposed that at some point you may want to have an energy storage device in everybody's home. It would allow you to economize on your own fuel consumption. Maybe it would go that way. That could be a real change in the models of how we think about generation consumption, getting down to the individual consumer. It very much parallels the discussions around self-generation of power.

Twenty years ago, if you had done a study or you had thought it through, it wouldn't have made sense. But now, you start getting renewables as a form of self-generation, solar panels on somebody's home, or community-scale wind. It changes this whole opportunity. You get generation devices that are smaller, microgenerators. When I first look at this as an engineer, I think about if it really makes sense, for example, for everybody to go to Home Depot and buy a generator. Is that going to allow us the most efficient use of our energy resources to make electricity?

At first glance, I think my engineering sense says "no way." That's really inefficient. But when you think about the transience of how you use electricity, when you go home and you start and stop the generation, or consider your use of electricity, your consumption of electricity, it affords a different model to think about how you're generating. Because if you can exactly match supply with demand, you'd gain a lot of efficiencies. So I wouldn't go to the extreme quite yet of saying that's the preferred package. Certainly when you go through the economics today, it is probably not the preferred path. But as our technology gets better and energy storage becomes cheaper and more reliable and scalable, both up and down as you are anticipating, I think the world could look really different in the future. It may get down to the individual level.

Stern: One day, will we have a backpack with one of these devices to walk around with? To power all our smartphones, computers, and devices?

Merfeld: You may. I will tell you, it won't look like the type of energy storage that we're producing today. It's not really how it was developed. It may be a derivative. It may be a permutation. It may be a completely different technology. And again, my approach is that it doesn't have to be what we're making today.

Stern: You've been doing this project for six or eight years, and you still have X-number of years to go. Do you see a timeline on this and/or do you see another project down the road for you?

Merfeld: Yeah. Those are the things that keep me excited, thinking about what's the next project down the road. That's the role of research in our company. It's our job is to anticipate what problem our business counterparts haven't thought of yet. How do we bring advanced sensors and better analytics to our manufacturing processes, so that we can make the energy storage even more affordable, make it even more reliable? There's a lot of fertile ground still in it, in this type of technology. In parallel, we have pipelines of activities that really turn this technology on its head. And these are technologies that are on timelines that will take us out three, five, and even ten years. We're not going to forget and lose sight of the fact that there's a lot we need to do still before we've got a really successful energy storage business with our first launch products. We'll stay focused on that.

But at the same time, as we see these other ideas, we're going to be challenged to pull them in sooner, to pull them in faster. I think the world is in a state today where technology is evolving not linearly, but exponentially. I think it's a power function of advancement, and that's exciting because as long as you have a mindset that you constantly are going to be putting yourself out of a job because you did a great job of solving today's problem, there is going to be plenty more to solve.

Stern: From this research that you are doing, is there a product on the marketplace right now?

Merfeld: Yeah, so there is a commercial product that's been launched this year. We're ramping up production in our facility here in Schenectady, New York.

Stern: What has been the response in the marketplace to the product?

Merfeld: It's been incredibly favorable. I think the challenge that we are having right now is making sure that we keep the pace of our abilities to ramp up the desired volume in check. What we don't want to do is get too far ahead of ourselves. We want to make sure that when we're shipping product, it's fully vetted, it's reliable, and it's gonna do everything that we promised. To make sure that we have a successful launch, I think we're putting on the brakes a bit in terms of pushing back on some of the customer demand.

Stern: Can you talk a little bit about the intellectual property side of the work that you do? I realize sometimes GE goes out and buys technology. But sometimes you develop it in-house. What is the process GE does to secure the rights to the ideas?

Merfeld: So mechanistically, we use every tool that's out there. Every means—and we are very aggressive about patenting. We also place a lot of emphasis on know-how. We do trade secrets. We have formalized systems for both of those obviously. We do a lot of work on disclosures. And from my perspective, a lot of the work that we put into disclosing our ideas in writing up patent claims, to me serves a purpose not only to protect the IP, but it allows us to take a step back and frame what we know. Many times in the process of doing that, we understand where our gaps are, and it gives us an opportunity to go back and do a better job.

You had mentioned, in terms of acquisition, we have no shame in realizing that within our company. We may not have the world's leading experts. We may not have all the skill sets that you would ideally like for a given project. So in those instances, and it was the truth in this battery technology, we were fortunate that we were working with a small group of engineers and scientists over in the UK ten years ago that had some principle scientists that had been working on this technology their entire careers. We made an acquisition where we brought them in and made them part of our team. And there's no substituting that career's worth of experience in getting to a good commercial solution. I think that is a good example of how we expanded our intellectual know-how, by bringing people to the table that have been thinking about this problem for a long time. And then we teamed them up with people that were relatively unbiased by history and who could question why things were done the way they were. It's an incredibly strong team.

Stern: Unrelated to the work that you're working on, do you have any favorite technologies or products, or nontechnologies that you just like having around where you work or where you live?

Merfeld: So certainly I am as culpable as anybody of enjoying my iPad and my iPhone. And I'm actually driving a hybrid vehicle right now as part of the fleet of test vehicles that we have here at our research center. I'm driving a Volt. I have been for the last two weeks and I think I get it for another week. I just think it's inspiring to be able to use this technology in a way that five years ago, we probably weren't even aware that we had the capability. So even though it doesn't directly relate to the projects that we are working on, I think it's inspiring.

For example, the interface that a lot of these new technologies have is a big component of how you utilize them. When you get in a Volt, a large portion of it is this interface that you have in the vehicle. So you understand how when you press on the accelerator, it influences how you consume electricity. When you have an iPad and you have these apps on there that are intuitive such that my three-year-old son can get through it, you know, I find those sorts of things inspiring. I think there's a big component of human interface that we've got at our fingertips today that are gonna end up migrating even into these industrial lines–type technologies that we've been talking about.

Stern: Do you have any advice for would-be inventors or people who are out in the field today?

Merfeld: Number one, I think you've got to love what you're doing. If you have the luxury to define your role, foremost make it about doing something that gets you excited. I feel fortunate because I've got a job that allows me to craft it and tailor it into something that keeps me excited. So it's not about the money, it's not about the glory. It's about staying challenged. I think engineers who aspire to solve real problems are what we need. And I'd love to see more of those folks coming out of school and creating new technologies that I can utilize.

Stern: Are there any particular fields of study that you recommend people to look into?

Merfeld: Absolutely. I think material science and chemical engineering are just an incredibly solid foundation to have to go after any of the problems that we talk about. I see those skill sets being applied almost universally across the problem sets that we see. I think that's an area where those folks are valued in industrial research and general research, so hopefully they will continue to find profitable jobs that pay them well. I think that's going to inspire people more and more to continue to do research in those areas. I think that the interfaces between different sciences are always going to be ripe for new innovations. You see that happening at the interface between biology and material science. You can almost take a pair of dice and put all the fields of sciences on the face of those dice, roll them, and think about how you could develop a career that lands anywhere between the interfaces.

Stern: When you're not working, what do you do for fun or distraction?

Merfeld: I'm fortunate. I'm married to an electrical engineer who is very much at the forefront of driving GE's solar technology. So we have three children together and that consumes a lot of our time.

Stern: When you're at a dinner party with your wife, what do you tell people that you do?

Merfeld: I tell them that she leads up the efforts for GE on the solar front and I'm tied in with leading up what we do on the energy storage front. So people always ask us how do you put solar and batteries together? And we always tell them that's something that we're working on.

Stern: At some point, do you ever plan to retire? What do you think you'll do?

Merfeld: I don't think my wife or I are either the type that would ever be aspiring to sit around on a beach doing nothing. I think we're going to have long careers doing innovation. Sometimes we like to daydream about what our second career would be. If I could do anything, I think the medical sciences are fascinating. I've always thought about tailoring some of my background to be more involved in physiology and medicine. My neighbor, who is a surgeon, inspires part of that. I think that interface is interesting.

The other side that would be more back to my roots from the farm. I adore doing projects around the home. I think it would be terrific to be a fine craftsman who can—without limitations on the project budget—do terrific things to develop landscapes and woodwork. I enjoy stuff like that.

Steve Gass

President
SawStop, LLC

Steve Gass *was born in Portland, Oregon. He has an undergraduate degree in physics from Oregon State University, a PhD in physics from the University of California, San Diego, and law degree from the University of California, Berkley. Upon graduation, he became a practicing patent attorney, where he conducted patent prosecution and litigation in a variety of technologies, including medical devices, computer technology, and woodworking equipment.*

Since 2000, he has been president of SawStop, LLC, where he invented and patented the technology for a table-saw safety system. He is the principal inventor on over 80 patents.

Gass is a lifelong woodworker and enjoys skiing, kayaking, and whitewater rafting.

Brett Stern: Please tell me about your background. Where were you born? What was your education and your field of study?

Steve Gass: I was born in Portland, Oregon, and grew up on a horse ranch out in eastern Oregon in a small town called Hermiston. Then I went to Oregon State—undergraduate—studied physics, and graduated in 1986. Then I went to the University of California, San Diego, also studying physics. I got my PhD in 1990. My thesis was on protein folding dynamics, so a different field than what I am doing now.

About halfway through my graduate work, I realized that academic physics probably was not going to be for me. I always thought I was going to do that—just be a professor, but I came to realize that you spend a lot of time writing grants as a professor, and I really just didn't have much interest in doing that. So I looked around for what else I could do.

A friend of a friend mentioned patent law. I didn't know anything about it. I talked to a patent attorney who described the job and said it was interesting and paid well, and you got to be involved with technology. I've always loved building things and figuring out how things work, so I definitely wanted to do something that involved applying engineering technology. I decided to go to law school and become a patent attorney, and so I did that.

I graduated with my PhD in 1990, and then went to law school. I started in 1991 and graduated in 1994 from the University of California at Berkley. I then came back to Portland and I went to work for a patent law firm, here in town. It's a boutique patent firm. All they have is patent attorneys. I worked there from 1994 through 2000. I became a partner, and then the last year I was there, I had the idea for SawStop.

Stern: Growing up, would you consider yourself an inventor, always building things or making things?

Gass: I was definitely a tinkerer at the very least. I always loved Legos and built things. My whole childhood was about building things, whether with a fort or a toy or remodeling a van. That's what I did with my time. Always building something.

Stern: Any particular invention that you remember growing up?

Gass: I don't know that I could put my finger on one in particular. On the farm, there was always something that you had to figure out or get to make work. But, it was just such a normal part of growing up, doing stuff like that. None of it particularly stood out.

Stern: Could you explain how SawStop works?

Gass: The way the system works is that we are able to detect contact between the operator of a table saw and the blade, and distinguish that from contact between the wood and the blade. We electrically insulate the arbor shaft that the blade is mounted to, and then we apply a signal, a 500 kilohertz sine wave, about three volts capacity coupled on the blade. Your body has a capacitance of roughly 100 picofarads, and the blade has a capacitance to ground given the insulation that has been put in there, of about 100 picofarads as well. So when you touch the blade, you essentially double the capacitance of the blade and that drops the voltage on the blade, by let's say a factor of two. So you go from roughly three volts to less than half.

There is a microcontroller in the saw that constantly monitors that signal and when it sees it drop, it knows that something other than a board has come into contact with blade. It then triggers a very high-speed brake, which is a chunk of aluminum— very soft aluminum—that gets shoved right into the teeth of the saw blade.

The actuator that does that is a spring, about one hundred fifty pounds that is compressed. In its compressed state, it is held back by a tiny little stainless steel fuse wire. There's a capacitor that stores a charge. When the system detects contact, it then releases that charge to melt, essentially vaporize, a very small segment of the fuse wire that severs it and then releases the brake to push out into the blade.

That's the more technical side. If you want it from a more layperson's side, the blade becomes a sensor a lot like a touch lamp. So when a user contacts it, we are able to detect that and then the brake gets shoved into the teeth of the blade and stops it in just a few thousandths of a second. For context, that is probably about ten times faster than your air bag would deploy.

Stern: Would you say that your background in physics gave you this direction?

Gass: Well, it certainly enabled it. In some ways, the physics in the basic system is really high school-level physics, but you know what happened is when I had the idea originally, it was a question. "Hey could I stop that blade? I wonder if I could stop that blade fast enough that if I ran my hand into it I wouldn't get a serious injury?" The physics enabled me to answer that question. Because you might think that the first thought would be, is it within the realm of physical possibility? Does it take ten million pounds to stop the blade fast enough and how fast do you have to stop the blade? I just stepped through the process of analyzing that question.

The first part of it is, "Well, how fast do I have to stop the blade?" And I thought, "Well, how deep a cut am I willing to tolerate," assuming that I can't start to stop it until I've already contacted the blade? My initial thought was probably one-eighth of an inch is about as deep as would be practical for a normal to slightly high-speed injury. I figured, "If a high-speed feed on a table saw would be on the order of a one foot per second. It would be fast, but not out of the realm of ordinary to feed material at that speed. It's fairly common for people to literally feed their fingers through the blade just as they're feeding material in. So I figure about one-eighth of an inch at one foot per second. Well, one-eighth inch is about one one-hundredth of a foot. That means I've got to stop the blade in about one-hundredth of a second so that my finger doesn't travel more than one-eighth of an inch in the time it takes the blade to stop.

And then I calculated the blade, which weighs about two pounds, and figured out how much force it would take to stop that blade, assuming that I applied the force on the outside of the blade, in ten milliseconds. That turned out to be on the order of one thousand pounds, which seemed doable. It wasn't one million pounds, which probably you wouldn't be able to do. So I thought, "Well, that seems viable."

Based on that analysis, I didn't think I'd have any problems stopping the blade. Then the question was, "Is there some way I could tell the difference between the wood and the person?" That was a thought process of, "Well, what physical differences are there between wood and people?" Wood is hard, so would you have some vibration on the blade that would not be absorbed by the hard wood but would be absorbed by the soft flesh? You'd be able to detect that. I thought about an optical system to look for blood on the teeth of the blade and pass through.

Ultimately, I thought about of a proximity-based system for looking at a radio frequency type signal, where you put the signal on the blade and you look for the hand to get close to it. But from a physics standpoint, I thought that there's not going to be any good way to get a discreet signal change when that happens. It's going to be a very small change, very hard to detect, and subject to a lot of inaccuracy.

Ultimately, I came up with this contact detection that is like a touch lamp, and that seemed viable. I thought, I didn't have any physics reason why it wouldn't work. So then I built a prototype. The physics allowed me to have the tools to analyze it to determine whether it was plausible or not. Now, someone who didn't have that background might have had to take more of a chance in a sense that I didn't have to take. They maybe would have to hire an engineer and spend thousands of dollars to have it analyzed by someone else without having any certainty of whether there was some fundamental reason it wouldn't work. For me, I was able to do that just on a piece of paper in the shop without a lot of effort. It allowed me to go forward without having to invest a lot. And that, I think definitely facilitated the development of the invention.

Stern: You found this problem because you have a woodworking hobby. So you were in the shop, and you saw this problem?

Gass: Yes.

Stern: Your approach to this whole project is not from a corporate setting but from an individual point of view, where you were in your garage, which is where a lot of people think invention comes from. You saw a problem and then you used your knowledge, whether it was Yankee ingenuity or your PhD, to solve that problem.

Gass: Yeah, I think that most inventions, in my experience and certainly in my case, the inventions that I've had always come out of a problem that I see, in work I'm doing, in a tool I'm using. It's not an abstract, "Okay, where's there an economic opportunity out there that I can capitalize on?" And I'll invent something to address that. I'm doing the things I do in my ordinary life, day-to-day, and I see something that might be done better or that doesn't work the way I'd like it to. And I figure out a different or better way to do it.

Stern: But at that point, your day-to-day was as a patent attorney. So, you were looking at inventions all day and at the inventing process.

Gass: Yeah, definitely. I'd say I had the exposure on an everyday, basis.

Stern: The clients that you had, were they generally corporate clients or more individual inventors?

Stern: It was a mix. We represented both individuals who had an idea and just were looking to come in and protect it, and starting that process. And then also we had corporations that we represented: Mattel, Yakima Products, which makes roof racks for bikes and stuff. Also Acumed, which makes orthopedic implants.

Stern: In the table-saw market, what was the state of the art prior to your work? How did the industry solve this problem?

Gass: Well, it's interesting you describe it that way. I would say that the industry didn't see that they had a problem. It's an interesting analysis. The table saw causes a huge number of accidents every year. About sixty-five thousand accidents every year that require medical treatment.

Stern: And how many fingers are cut off?

Gass: About four thousand people experience amputations of one or more fingers. And at a huge cost. Over $2 billion a year in economic costs from those accidents. And to put that number in context, the total market for table saws, retail market, if you add up the price every table saw sold in the United States, it's probably on the order of $200 to $300 million. So you're talking about a product that is doing nearly ten times its purchase price in economic harm from injuries.

Then you think, "Wow, that would be a huge problem for the manufacturers." But in fact, it's not. Because the manufacturers don't pay for those injuries. The rest of us do. First of all, of course, and primarily, the victims pay for it. But then the rest of us pay through Worker's Comp, insurance costs, disability payments, reduced productivity, and all those things that society as a whole pays for that the manufacturers don't. So they really didn't see this as a problem.

The standard solution they offered was a guard. If you do woodworking, you know that the vast majorities of woodworkers don't use the guard. Especially what they used historically was such an awful contraption that was difficult to install and remove. And it had to be removed for certain operations. The saws in general were well made. The controls were easy to use. But you had this cheap, plastic guard sitting on top that was really just a piece of junk.

What I came to conclude was that the guard was placed on the saw not to protect users but to protect the manufacturers from product liability. They put this guard on there, and they say in the manual to always use the guard. Of course, most people don't use the guard. Manufacturers know that, but when someone has an accident, they usually don't have the guard on when the accident occurs. The manufacturer stands up in court and says, "Hey, you know, we told them to use the guard. They weren't using the guard. It's their own fault."

So it's a product feature that offers no benefit to the customer but is a great benefit to the manufacturer in terms of insulating them from product liability. Now as a woodworker, I knew the guards weren't used. I had never seen a guard with a table saw, and I never used a guard at a table saw prior to the time I started SawStop.

My first memories were in my dad's shop, and I've been a woodworker since. I grew up with a woodshop and table saw, so I knew that people didn't use the guard, and I knew that people had accidents on table saws. I didn't see the guard as a solution, and these accidents were certainly a serious problem.

Stern: Your day job was as a patent attorney, but at night or the weekend, you were working on the prototype of your invention?

Gass: Yeah. It took me about thirty days from starting, scratching my head about it, to having a working prototype.

Stern: How did you test your working prototypes?

Gass: Well, with an invention like SawStop, that's a bit of a challenge. Initially what I did was actually touch the side of the blade with my finger, because you don't get injured touching the side of the blade.

Stern: What was going through your head when you said, "Yes, I'm going to stick my finger into a moving table saw blade." Because your whole life you've been told to not do that?

Gass: Yeah. It definitely runs contrary to all your instincts. Your body is telling you, "No, don't do this. This is stupid."

Stern: Was anybody there with you?

Gass: Yeah, the first time I did it, one of my partners in crime here at SawStop. When I had the invention, I got three of my friends at work involved. It was a bunch of patent attorneys, and one of them was there with me when we were trying to capture the data for the first time. And it definitely is something you definitely don't want to do if you don't have to.

Stern: Can I ask if you are left- or right-handed and which finger, which hand did you push in?

Gass: Well, I'm right-handed. We went through a significant thought process to decide which finger. We decided definitely the left hand, and the best was the ring finger. Probably the best option because you use your index and middle fingers a lot, and your thumb. Your pinky is on the outside, so you are more likely to bump it, so figured the left ring finger was probably the most protected sacrificial victim for a cut.

Stern: Also, you always need your middle finger, no matter what.

Gass: Oh yeah! You can't have that injured.

Stern: Okay, so you tested it. The machine works as planned. Your buddies are there and what went through your head and what did you say to each other? Was there an aha! moment?

Gass: There were a couple aha! moments. The first one was actually not with my finger. Well, let's back up. The first one in a sense was with my finger, but the blade wasn't spinning. I had this idea that maybe I could tell the difference between the wood and flesh. I was pretty sure I could make the brake work. I didn't think there would be much problem with that. The real uncertainty for me was electrically. Was wood different enough from flesh [for something] to be able to detect the difference? I really didn't know that.

I did hobby electronics throughout my education. I played with electronics and knew how to use an oscilloscope at home in my shop. And so I built a little circuit to do the detection. I had a saw blade I'd laid on the counter in my shop. I had the circuit there connected actively to it. I had set the circuit up and connected it to the oscilloscope. Then I would take my finger and touch the saw blade, and you could see the signal on the oscilloscope drop. I thought, "Okay, well, that's good." I can tell that I'm detecting it. If I could set up a test to see the drop so that I could have something trigger based on that drop… but I didn't know what wood would do.

Still in the shop, I reach over, grab the nearest chunk of wood, and stick it on the saw blade. I see, sure enough, no change in the signal. Then I think, "Well, huh, how about if it's wet?" So I literally lick the end of the piece of wood and stick it back down on the saw blade, and no change. And, okay, what if it's, really wet? And so I take it to the sink and run it under the faucet, bring it back, and stick it on the saw blade. Still no change. And that really for me was the first aha! moment. It's gonna work.

I was able to have a fair amount of confidence at that point that I could make the system work. Then I went out and bought a saw to build an actual full-functioning prototype and brake for it. I used a little chunk of oak mounted on a bolt to pivot into the blade. The spring I used was a nose gear from a radio-controlled airplane, and the fuse wire I used was a small strand from a cable that was lying around.

Stern: So just stuff lying around the shop?

Gass: Yeah, it was all just stuff lying around the shop. Some capacitors from a power supply. Literally just duct tape and bailing wire.

Stern: Duct tape! You said the magic word.

Gass: Got the brake working on the saw, then turned the thing over, because I did most of the work initially upside down on the saw so I could see everything. Then I turned it over. I pretty quickly decided I wanted to use a hot dog for the sacrificial victim for the early test. I fed the hot dog into the blade, and it stopped, and there was just a little cut on the hot dog, and I thought, "I'll be damned. This will actually work." Those were the two biggest aha! moments, when we were doing the very first hot dog test, and then the first test with my finger just on the side of the blade on the counter.

Stern: Would you like to guess how much you spent, not including the hot dog, on these first prototypes?

Gass: The saw was by far the largest thing. I think it was $200 or $250. I'm sure the first prototype was under $300.

Stern: So the prototype of actual parts and pieces was maybe $50 to $75?

Gass: Yeah. I didn't buy anything to make the pieces. It was all stuff I had lying around the shop.

Stern: At what point did you start the IP process?

Gass: Immediately. In fact, I got one of my friends involved to do the first patent application with me. That was almost simultaneous with the process of developing the prototype. I hadn't finished the prototype. As I recall, the patent got finished at about the same time the prototype did. Clearly this was a significant component from the very beginning. Because as a patent attorney, I knew that unless you have protection for your invention, companies are not going to pay you for it, and you can never recoup anything from the investment you made to develop it.

Stern: Certainly, within the woodworking industry, there was no prior art. As you said, it was just a piece of plastic that covered the blade, a blade guard. Was there any other prior art that you thought you had to get around in other fields?

Gass: Not much. From a commercial product standpoint, there was never anything like it. There were a small number of patents out there that were similar in certain aspects, but none of them actually described a system that had worked. They were either too slow or they didn't have robust detection, or both. That was really unique from a patent perspective. That is a very unusual circumstance.

Typically what you are doing is improving on something else. So your patent protection is really in some way limited to the improvements you're making. You can't protect the whole solution. In this case, there was so little that had been done before that we felt like we had a good chance anyway at getting very functional protection to solve the problem. So we weren't just going to get protection for one way to do it—we thought we had the potential to get protection for the most viable ways to solve this problem.

Stern: So you had numerous claims?

Gass: Yes. First we filed a provisional patent application in October 1999. I don't know that we had any claims into that.

Stern: The provisional patent application is really to set a date in time?

Gass: It sets your disclosure from which you can then later make your claims, but it gets your date recorded. Then, as we went forward, I initially tried to get some manufacturers interested in it. They weren't interested.

Stern: Whom were you speaking to?

Gass: My primary contact at Delta was the vice president of engineering. They looked at it for a couple of months and then I got a call and he said, "Uh, you know, well, the marketing guys say safety doesn't sell, so we're not interested."

Stern: Were they looking at the actual prototype or the patents?

Gass: They were looking at the patent I'd written, which was a technical disclosure of what was in the prototype, how it worked, as well as a video of the prototype so they could see how it was constructed and how it worked. So they had a fairly complete disclosure of what was there.

Stern: Did you ever do a face-to-face with them and did they sign an NDA [nondisclosure agreement]?

Gass: They did sign an NDA at the very beginning. I did not have a face-to-face with them. There were maybe three or four companies that I then e-mailed or wrote a letter to, and ended up getting enough interest to get an NDA. Oftentimes, if you can get enough interest from somebody to sign an NDA, that is a good sign. It indicates that they are interested in whatever it is you've got.

Stern: Do you feel that you have any advantage being a patent attorney in that you can talk the talk?

Gass: Yes, because I'm familiar with how to protect the invention and the NDA. If you approach Black & Decker, they will make you sign sort of an anti-NDA. It's the opposite of an NDA. It says they can do whatever they want with it. And so, I didn't go to Black & Decker. I could look at the contracts and see what the implications were, and what was reasonable and what wasn't.

Stern: You had fellow patent attorneys partner up with you on this project from the beginning?

Gass: I did. I didn't pick them principally because they were patent attorneys, but really because they were my friends.

Stern: What was the reason for having friends participate in a project like this? Were they in the shop with you tinkering around?

Gass: Not as much that. It really was more that I was excited about it and sharing it with my friends at work. Then what happened is that when the original manufacturers we contacted and all indicated they weren't interested one way or another, we decided that was kind of crazy. We thought woodworkers would be interested in this and it was a good thing.

We thought, "Well, what can we do? We can just let it go, or we can make a nicer prototype and take it to a trade show and make it public." We decided to do that. We hired an engineering firm. That process we knew was going to be expensive, both hiring the engineering firm and going to the trade show. We figured it would take about $150,000 to do that, more money than I had available.

There was also a lot of legwork that needed to be done. Just to get ready to do it and managing it was more than I could handle on my own. I also knew there would be more patent work to do.

Stern: How long was that—from the point you discovered the woodworking machinery companies weren't interested to the point of deciding to hire engineering firms?

Gass: In some ways there was some overlap. October 1999 is when I contacted Delta, late October 1999. I think they told me in February 2000 that they weren't interested. But, as the fall wore on, I was calling them maybe once a month saying, "Hey, what's up?" And not getting much back. They had just moved their headquarters and said, "Oh, we're going to need some more time." So they were delaying, but it wasn't clear that they were interested.

Ultimately they told me—there was a black-and-white statement of a lack of interest, but the indications were in that direction already. When they finally said no in February, we had already planned that we were going to hire the engineering company and move forward on building the prototypes just on the theory that they weren't interested.

Stern: Building the prototypes and hiring the engineering firm was with the incentive to do what next?

Gass: To make it public. To have something you could demonstrate at a trade show to generate consumer interest. First to have people see it and then they go tell Delta, "Hey, I want that feature on my next saw. You should do that." And tell the other manufacturers the same thing.

We also figured that at the trade show, we would be able to show it directly and have an easy face-to-face meeting with a bunch of different manufacturers, because they would all be at this trade show. It was a woodworking trade show that they all attend. It was something that allowed us in one fell swoop to potentially expose the invention to a large number of different manufacturers.

Stern: What was the response from the general public and the response from the industry?

Gass: From the general public, it was really overwhelming. Because we were kind of late signing up for the trade show, it was in August 2000, we had only six months to build the prototype and get into the trade show. We got a booth that was way off the main floor in a conference room on the third floor off to the side, really a lousy location from a booth standpoint. But when we started doing the demonstrations, word got around. It's a very impressive demonstration to see.

Word got around, and pretty soon people were just jammed shoulder to shoulder in our little ten-by-ten booth, in the aisle around our ten-by-ten booth, in the booths surrounding the aisle around our ten-by-ten booth. The demonstrations were packed to point we couldn't get more people in to see them. One of the woodworking journalists pulled me aside and said, "You know, it's like you guys are giving away one hundred–dollar bills at the demonstration." It was really overwhelming. We had no idea we would get that much attention. We were pleased, obviously, but really had no expectation of that going into it.

Stern: And from the industry?

Gass: From the industry, there was a lot of interest. People would say it looked like we were the DeWalt booth at times because there were more guys with DeWalt shirts in our booth than there were SawStop shirts. A guy had a picture of the president of the DeWalt brand there standing shoulder to shoulder with this guy from Delta looking down at our saw and talking to us. It really was effective in terms of getting the attention from the manufacturers.

Stern: Six months later, what was the outcome of all of that?

Gass: Nothing. Six months later we were still having discussions. What happened is we had the trade show, we get everybody's attention, we get a series of meetings with the manufacturers, and they are all interested. They all want to talk to us. We fly to all the different manufacturers to show them our prototypes and talk with them about licensing and what we want. The challenge for us is that the invention really is a neat thing and everybody likes it and can see the benefit from it, from the people-not-cutting-their-finger-off standpoint. But from a financial standpoint, none of the manufacturers can figure out how to make more money doing it than not doing it.

Stern: They're not going to sell more table saws.

Gass: They're not going to sell more table saws, or at least, not if everybody has it. And we felt like it was a technology that we couldn't license exclusively to one manufacturer, from an ethical standpoint. So essentially what we were saying is you guys need to do this because it's the right thing to do and if you don't do it, you're going to get sued because people are going to get their fingers cut off, and they're going to know there was some way you could have prevented it, and you'll be responsible for that.

Stern: Were companies looking for exclusivity?

Gass: They mentioned it, but even if we would have given exclusivity, I don't know that it would have made any difference because none of them wanted to do it on their whole product line. I think they wanted to make one product and offer it. And that's great, but they were also afraid of the secondary ramifications of proving that the technology works.

Because they were in the situation where there wasn't anything better than the guard out there, and so when someone was injured, they went into court and said, "Hey, the guard's the best thing out there, and there's nothing else. It's not like a saw can tell a difference. So very sorry, but it's not our fault the guy got injured." And that was effective. They never lost these cases. But then this technology shows up and they still can argue, "Hey, well it's not viable. This doesn't really work. It's too expensive. It's not reliable." That's what they have argued, in fact, for the last decade in these lawsuits.

But it's very different if somebody actually puts a product on the market and shows that it works. So they were in the position of, "Well, it would be interesting to do this on one product, but, man, if we do it on one product we're hosed for the rest of our product line, because we had just proven that this technology works. And once we do that, if we don't do it on our whole product line, we're going to get sued and lose." It was like a Pandora's box for them.

Stern: When you went to the trade show, it was with the idea that you would license the technology. Then, six months later, you realized that there was interest but really feigning interest. What was the next step that you decided to do?

Gass: Well, it wasn't six months. It probably took us almost two years to work our way through all the different manufacturers, and the licensing negotiations and discussions, and conclude that, in fact, this wasn't going to happen. We were not going to license it.

Stern: And you have money going out the whole time as well?

Gass: Yeah, definitely. And three of us quit our jobs as patent attorneys after we got back from the trade show and started doing SawStop full time. We got investment—friends and family investment. In fact, I mentioned that Acumed was one of my clients. The owner of that company invested some money. The father-in-law of one of my colleagues, David Fanning, invested some money. And that was enough. We took small salaries and we didn't have huge expenses. That was enough to get us by for a couple of years. Then when we realized that they weren't going to license it, it was like, "Okay, do we go back and practice patent law?"

Stern: You all had legal backgrounds and some technical backgrounds, but you didn't have the business background?

Gass: Yeah, and we didn't have technical backgrounds in terms of ever developing a product. I mean we had degrees, I have a PhD in physics. David Fulmer had a bachelor's in electrical engineering. David Fanning has a bachelor's in physics. David Fulmer worked in a lab, a semiconductor lab, and had electrical engineering experience making semiconductors.

None of us knew how to make a product. None of us knew how to use CAD software. We were just really completely blind from a product development standpoint. We had the audacity to say, "Well, heck, we think we can make a saw. Why not? How hard can it be?" But, we knew we had to raise money. We figured we needed a couple million dollars to do the development, tooling, and bringing it to market.

Towards the end of 2002, we wrote up a business plan proposal and started trying to raise money. We were fortunate that the invention, in a lot of ways, sold itself. The benefits are very easy to describe and recognize. You see the little video of the hot dog running into the blade and you get it. It is a pretty cool thing.

Stern: Did the money come from the venture capital industry or just people with money?

Gass: It was really again friends and family, and maybe a little broader—friends of friends. I had a friend who is an investment advisor. He invested some, and he had some other people that he knew that invested some. It was just a process of finding people that were basically all friends or friends of friends. There was no formal venture capital.

Stern: What percentage of these people do you think had an interest in woodworking?

Gass: I don't know that any of them were woodworkers as such. No—there was one. There was a doctor from Louisiana who had contacted us when we were at the trade show originally and said he wanted to invest if we ever had an investment opportunity. We didn't think anything of it at the time, and then when we finally starting raising money, we called him back. He wanted to invest, and he was a woodworker. In fact, I think his father had had a bad accident, which raised his interest in particular.

Stern: So you raised this money, and then you decided to tool up and go into production?

Gass: We designed the saw. We designed the electronics or hired the people to design the electronics for us. We contacted manufacturers in Taiwan that build these types of products for US manufacturers. It's a little misleading to call, say Black & Decker, a manufacturer. By and large, they are a brand that contracts with OEMs [original equipment manufacturers] in Taiwan. The factory we worked with in Taiwan, there's a production line set up for SawStop, and right next to it is a Black & Decker production line. Most of the table saws and woodworking equipment have historically have been made in Taichung City in Taiwan.

At that point, we were three guys operating out of the top of my barn. We set up production of a saw without having to rent a plant and figure out how source stuff because they had all that there [in Taiwan]. We just went over there and wrote a check for the tooling, and they shipped us a container of saws. Now it's a little more complicated than that, but in essence, that is what happened.

Stern: So now you had a container of table saws. How did you go about marketing it?

Gass: We knew in advance we were going to do this, and we started taking what we called preorders. We were still going to trade shows to publicize the technology, and we started taking preorders when we decided to we were going to build saws. We had five hundred people on our preorder list when we started, when the containers came in. We just started calling the people on the preorder list. We were able to basically jump-start the business by filling those first pre-orders. And we were cash flow positive really from the time we started selling saws.

Stern: Were they generally being sold to hobby woodworkers or job shops?

Gass: They were job shops initially. And really, that's largely because of the nature of the first saw we sold. It was an industrial saw. Some people would buy it for home use, but it was an expensive and big saw, so it was really targeted more for industrial use, and that's what most of our initial customers were. In part, because we were going to the industrial trade shows. We had a web site, and people would turn in preorders on our web site. We had a surprisingly high percentage of people who turned in preorders and, in fact, then bought the saws, which gave us an initial market.

We started advertising, and then we set up dealerships throughout the country. Really, the three of us that started the company, we're really more engineering types. We're not people people. We're not good at marketing. We are not good at sales. That's just not our thing. We knew we needed to bring in somebody who liked to do that kind of thing, be good at it, because we weren't going to be.

So we brought in Paul Carter, who was actually a friend of mine from high school that I hadn't talked to in years. He saw some article in the paper or something and called me. He was working in Wilsonville, Oregon, at the time, at In Focus. And he said, "Hey Steve. Paul Carter here. You know, I saw your thing. Let's have lunch. Love to hear about it." We ended up hiring him in 2005. We started selling saws in late 2004, and he joined us shortly thereafter in February 2005, and took over the marketing and really built up that side of the business.

Stern: If you can jump forward five or six years, where is the company now and what is the response from the industry?

Gass: We've taken over the industrial table-saw market. We're the number-one selling saw, I think by a pretty fair margin at this point. We've really decimated the rest of the cabinet-saw market because when spending that much on a saw, it's not much more to get one with SawStop. Most people like their fingers, so most people are buying our saws. Then we expanded our product line, so we have lower-cost saws as well. We're continuing in that direction, but right now, our cheapest saw is a contractor's saw at about $1,600.

Stern: Something that is portable?

Gass: More portable. It's a bit of a misnomer in that it was originally made to be portable, but in fact most people who buy it now use it in their garage. But, it still weighs a couple hundred pounds. The next saw we're coming out with is what's called a job-site saw, and it will be portable. Like one guy can pick it up, you know, it will be about seventy pounds or something like that.

Stern: What has been the response from the industry?

Gass: They hate us. They all got together and formed a consortium to try and find a way around our patents. They formed this consortium and they voted on how to respond to us to make sure they were all on the same page. In fact, my personal belief is that they used that to avoid licensing it to make sure that no one got off the reservation, so to speak. They continued to coordinate their activities to try and oppose adoption of the technology to make sure it didn't happen.

We continued to push, not only in the marketplace selling our saws, but we filed a petition at the Consumer Product Safety Commission to try and get them to create a rule that would require all table saws to have some active injury mitigation technology. That process has moved forward at a glacial pace, but in October of 2011, the commission voted to take the first significant step toward making the rule, and they voted five to nothing in favor of doing that.

That sent a huge shockwave through the industry. I think that indicated that this is going to happen. Particularly, in the fact that it was unanimous. When you actually look at the statistics of the injuries and the cost of those, and the cost of implanting something to prevent, the payback is somewhere between ten and one hundred—to one. It just makes so much sense. And even the Republicans who are generally more antiregulation were in favor of regulation in this particular area.

Stern: During this period, have you had dialog with the industry?

Gass: We have chats with them on occasion, and I've been deposed about a dozen times. I was testifying in a case against Black & Decker last week where someone had been injured. We end up having on-and-off discussions. We really haven't had much in the way of serious licensing discussions, I would say.

Stern: It seems like with all the effort that they've done to organize this trade association and all the effort they've put into it, they've spent more than they would have paid to you.

Gass: You know, that's a tough question. It's interesting. What they've spent on development through their trade organization, they've developed this alternative technology that uses an explosive retraction rather than braking the blade, stopping the blade that is. In some of the litigation, we've seen some of the cost. They spent less on that than we've spent on our R&D, so they try and trot that out as if they care about safety.

And the whole industry got together—all these multibillion dollar corporations—and pooled their resources to bring the best minds in the industry together to develop a solution to this terrible problem. But they spent a few million dollars, you know, and we spend a few million dollars *a year* on R&D. It was a very disingenuous effort. I think they were largely doing it to facilitate the coordination of their efforts and to then offer some shield in litigation to be able to say, "We are trying."

Stern: Originally, you started this project because you saw a need and you solved the problem. Now you are working on the business side. Do you feel amiss that you can't be out there every day looking for more problems to solve, or does the business create problems to solve?

Gass: The business creates plenty of problems to solve. The problem with it from my standpoint is that I don't like solving those problems. I enjoy the technical problems. I wake up in the morning, I get in the shower, and that's what I'm thinking about. I stand there and I'm thinking about how to make this spring connect to this lever. I hate having to deal with employees. When we had to do an employee handbook, it's like "Oh God! No!" I'd rather have a root canal. The sales side of it is the same thing. I just don't like those kinds of things. It's not what makes me excited. But you have to deal with it to have a business. So we do deal with it, but it's not something that feeds me very much.

Stern: Could you talk about where you seek and find inspiration or solutions?

Gass: It depends on what the nature of the problem is. From a big picture standpoint, I get a lot of inspiration from the e-mails and contact I have with customers. It's a pretty darn cool thing to talk to somebody who's run their hand in the blade and had a minor injury instead of having their hand mangled. I mean, that really means a lot to me. I'm not a money-motivated person. I was making more money as a patent attorney than I've made doing SawStop.

SawStop's been a lot more fun for me, but it's not that I've made more money doing it. It's just not that important because I make money either way. I like to whitewater kayak, and I can afford a new kayak now and then. I don't need that much more.

Stern: How has the success of the invention changed your life then?

Gass: I would say it's the idea that ate my life. I had a very different life before I started SawStop and I never anticipated it would end up being what it is. I ended up quitting my job, changing careers. It being a full-time endeavor for the better part of my life now. I had no idea that this [would come about] when I first had the idea. I was thinking, "Oh, this is cool. You know, I'll license it for a small fee and maybe I'll make a little bit extra money doing that." But I didn't really think of it as a change-my-life invention originally. It just wasn't what I visualized.

Stern: I think a lot of individual inventors think they are going to come up with an idea, get a patent, and just license it to a big company. Does that ever really happen in the real world?

Gass: Very, very, very rarely. One of things that most inventors—and I can say this from professional experience as well as my own personal experience with invention—[don't realize is that] getting the patent and solving the technical problems are the easy part. It's the marketing and finding somebody who can get it into a company and sell it to a company, that's the hard part. Inventors often don't have that skill set. They're engineers or technical by nature, and you really need a partner for the business side that has that other skill set.

Stern: Do you have any advice for would-be inventors?

Gass: Find somebody that has that deal-making skill set. I don't know how you find them. That's the hard part. A lot of is luck. Finding the right person who really knows how to do that, who likes doing that kind of thing. It's an interesting distinction in personality. I'm not gonna wake up in the morning, get in the shower, and think about who I'm going to call at what company and how I'm gonna pitch it. That's doesn't excite me, and so I'm not gonna be good at that. That kind of personality of making the deal and closing it and getting the signature—that's often a different personality than the inventor personality.

Even SawStop, which is, as inventions go, about as easy to sell as you can get. It really offers a clear advantage. But our pitch to people was basically, "You have to do this." That's not a good pitch. If we had had someone who was a better pitchman to start with, we might have had a very different outcome from the get-go.

Stern: Can you talk about your inventing process? Are you there with a pencil and the back of a napkin? Do you start sketching? Do you do 3D modeling? What is the process you go through to think of solutions?

Gass: I'm a big-picture person. I come up with a different way of approaching the problem from a schematic standpoint. I'm not as good at the details in figuring out where every screw and nut goes. So in terms of what I do, we have ten engineers here, and I'm noted for bringing a Post-It note in with a very simple pencil sketch and saying, "How about this?" Or, "Why don't you do it this way?" And then they draw it up in 3D CAD and show it to me, and we kind of go back and forth. But I'm definitely that back-of-the-napkin guy.

Stern: When you're dealing with your staff, do you have any particular method for motivation?

Gass: No. That's one of those drawbacks—I'm not a people person. That's not my strength, and we suffer for that. I think we're all aware of that. The three of us that I mentioned that run the business, one of our drawbacks is that we have a hard time being involved enough with our engineers to motivate them. It really is one of the things we debate and struggle with. How could we do this better? It's just something we can't bring ourselves to do in the way we think it should be done.

Stern: Is there a field of use that you are looking at applying your technology? Something that hasn't been done yet?

Gass: From a business standpoint, we are looking at expanding into other woodworking tools. Smaller table saws, more consumer products, and then handheld circular saws and miter saws. Really expanding out to have a full line of woodworking tools with the technology. We've talked to people who are using big industrial press rolls where you could [employ the technology so that if] somebody touched the roll it would stop. Same with chain saws, lawnmowers, sewing machines, paper shredders. It really runs the gamut of me thinking through, "Okay, is there some way I can tell if a person touches a dangerous area of this product?"

Stern: Has the commercialization side changed your life at all?

Gass: Well, certainly it did in quitting my job and what I do on a day-to-day basis. And it changed once we started actually building saws and had the business. I end up being drawn away from the engineering oftentimes to deal with business issues. I end up getting deposed a lot in lawsuits. Then I end up trying to sort of deal with the Consumer Product Safety Commission, trying to push them forward. That takes a lot of my time, which I would really personally rather be spending doing engineering. That and product development are what I like doing. You end up not having the time to dedicate to that, and that's frustrating as an engineer.

Stern: When you're not doing this, what do you do for fun or distraction?

Gass: Well, I like to get on my bulldozer and push dirt around on my property. I have a farm and I like riding the tractor. For more exciting fun, I do whitewater kayaking. And that's really my release. One of the nice things about it is, no matter what's going on elsewhere in my life, if I'm going down through the rapids, it's really very survival focused. You're not thinking about any problems in your day-to-day life. You're thinking about how you're going to keep your head above water and keep breathing till you get to the bottom of this rapid. And I really enjoy that kind of release.

Stern: When you're at a dinner party, what do you tell people that you do?

Gass: I alternate between saying I'm an inventor, and I run a business. It always has sounded funny to me to say I'm an inventor. Although I think that's probably the truest answer to what I do.

Stern: Do you plan to retire from this?

Gass: You know, I hope not. I would like to not have to deal with the business side of it, but I love the engineering side of it. It's what I would do if I won the lottery. I would still do that kind of engineering whether it was in this context or in some other. That's what I want to do with my days. I like thinking about how to solve problems. That's fun for me. That excites me. You know, it makes me want to get up in the morning.

Stern: Overall, would you say there are any particular skill sets that an inventor needs?

Gass: It's almost mostly a way of thinking. It's a willingness to challenge the status quo. To say, "Is there a better way?" To look at something and be willing to challenge it. I think many people just go through life and if something doesn't work perfectly for them, they think that's just the way it works and accept it. I think one of the keys to being an inventor is challenging that and saying, "Can I do this differently? I don't like the way that worked." In recognizing it and then challenging yourself to find a different way to do it.

Stern: Could you tell me your biggest failure and how you've overcome it?

Gass: I would say the licensing. We spent a couple of years, really in some sense wasted a couple of years, trying to license a technology. We overcame it by accepting that it wasn't going to work. One of the things that I think we've done well is to continue to be flexible as we go forward. We didn't make a plan and stick to it no matter what. Given what we know today versus what we knew six months ago, are we still doing the right thing? Should we be pushing in a different area? Should we switch our focus from this product to a different product? We dedicated a huge effort to that with licensing, and it failed. But we still said, "Okay, that's not going to work. What else should we do going forward?"

Stern: How would you say the technology and/or the business has influenced society?

Gass: Well, in our case we've had over fifteen hundred cases where people have run their finger into the blade and come away with a minor injury. If they were using any other saw, many of those would have been devastating, life-changing injuries. So we've saved a lot of people from serious injury. From a big-picture standpoint, I think we've precipitated a change that's in progress—where these kinds of products won't maim people in the future.

I don't know whether that's going to be five years from now or fifteen years from now, but it's inevitable that saws will stop hurting people. And fifty years from now, people will look back on it, and tell their grandchildren, "If you ran your hand in the blade of a table saw, it didn't stop like that one there." And, the grandchildren will be shocked that people ever used such a thing. I think that really will change society in that way.

Stern: Any final words or advice you want to give to an inventor?

Gass: Keep trying. Really, if at first you don't succeed, try something else.

Karen Swider-Lyons

Head, Alternative Energy Section, Chemistry Division, Naval Research Laboratory

Dr. Karen Swider-Lyons is head of the Alternative Energy Section in the Chemistry Division at the Naval Research Laboratory (NRL) in Washington DC. She earned her PhD in materials science and engineering from the University of Pennsylvania in 1992 for studies on high-temperature fuel cells.

She currently leads research programs on advanced battery materials, low-cost catalysts for polymer fuel cells, and the use of fuel cells for long-endurance, energy-efficient unmanned air and undersea vehicles. She has served as a technical advisor to the Defense Advanced Research Projects Agency (DARPA) and the Office of Naval Research (ONR). In 2010, she received the Dr. Delores M. Etter Top Scientist Award from the US Navy for her work on the Ion Tiger, a long-endurance hydrogen-powered fuel cell system for unmanned air vehicles (UAVs). Dr. Swider-Lyons has authored more than 80 technical publications and holds 12 patents.

Brett Stern: Can you give me background, your education, and your field of study?

Karen Swider-Lyons: I went to Haverford College for chemistry and to the University of Pennsylvania for graduate studies in materials science.

Stern: Would you say growing up you were an inventive child?

Swider-Lyons: No, but I always liked to do projects.

Stern: What types of projects?

Swider-Lyons: Well—frankly, to be sexist, as a girl—when I was left to myself, I did a lot of sewing and making clothing. As a teenager in college, I decided to refurbish my car, which I didn't know anything about. I just tried to get out there and do it.

Stern: In layperson's terms, can you talk about your field of study and the inventions that you work on?

Swider-Lyons: I work on electrochemical technology. Everything I'm working on now has to do with fuel cells and batteries, or derivatives of them. Because I work for the military, they are generally used in military systems. Sometimes, they're just general improvements to technology. So, the applications of my work span everything from consumer use all the way to military systems.

Stern: Do the problems come to you, or do you go out into the marketplace and look for the problems?

Swider-Lyons: We go look for the problems, absolutely. For instance my main focus now is on long endurance unmanned vehicles. There's a general need for longer endurance in unmanned vehicles. There's no current solution, and there's acute awareness of the problem.

Stern: Does someone in the military come to you and say, "Think about this"?

Swider-Lyons: Yes. I go to conferences and people bring up their needs for technology improvements: more efficient, cheaper, lighter. But there is usually no solution provided.

Stern: What was the state of the art in UAV [unmanned air vehicle] propulsion at the start of your work?

Swider-Lyons: I want to be clear that I am focusing on what is called small unmanned air vehicles—from about 15 to 100 pounds. These are powered by batteries and combustion engines. Batteries don't have enough endurance. The engines with heavy fuel offer better endurance, but they are very inefficient and unreliable—so both are inadequate for long-range military reconnaissance and surveillance. Traditional batteries can only sustain flight for a couple of hours. Traditional combustion engines can sustain flight for up to 24 hours, but they are noisy and emit a lot of infrared energy.

Stern: Could you describe the particular work that you're doing now?

Swider-Lyons: In 2009, my team flew an experimental but practical fuel-cell UAV, the Ion Tiger, for twenty-four hours with a five-pound payload using compressed hydrogen. Now we are gearing up for demonstrating about three days of flight with liquid hydrogen fuel. These experiments could revolutionize electric flight. Everyone wants to move to electric flight. I remember going to a meeting in 2004 at which people were talking about autonomous systems.

There was a big push, especially for autonomous airplanes. They wanted to take the pilots out, mainly just to get them out of harm's way. Autonomous planes have to get to the location and they have to send information back. You need a lot of endurance to do that. People came to me because they knew I worked on energy systems and they said, "How are you going to do that?"

I was like, "I haven't the foggiest idea."

With a colleague of mine, I was working on a supporting program of fuel cells for portable power projects. We'd go out to companies to visit their labs. One day my colleague, Bob Nowak, came to a meeting and held up a small black piece of plastic, and said "Look at this little fuel cell that this company, Protonex Technology Corporation, is building." I thought, "Oh. I wonder what we can do with that?"

An important invention a few years earlier from NRL was based around a two-gram video camera and what sort of airplane we could wrap around that. From this, they built, the Marine Corps' Dragon Eye, was a small battery-powered UAV in 2001. This was the first portable drone and is now on display in the Smithsonian National Air and Space Museum. Rich Foch is the lead inventor on that project. I was not part of that project.

We now had a really small fuel cell from Protonex. The Protonex fuel cell was about 100 Watts—equivalent of the power need for an incandescent light bulb—and it needed to run on hydrogen. The military had been focusing on 25-Watt fuel cells for portable computers, and also wanted them to run on heavy fuel, like diesel. Coincidentally, around that time my post doctoral fellow, Peter Bouwman, was bugging me to put a fuel cell on an airplane and fly it around. Until I saw that Protonex fuel cell, I hadn't thought it was possible because all of the fuel cells I had been seeing were too big and low power. And that's when we got the idea to do the first little fuel-cell-powered UAV, which was called Spider-Lion. Because we didn't have much of a research budget, we didn't have funding for reforming fuels or anything. We just carried compressed hydrogen in a paint ball cylinder—which resulted in us building a relatively simple system that worked quite well. Spider-Lion was just a six-pound toy to see if a new technology could work. Innovations are based upon such miniaturized experiments.

We were then able to fly it and people were, "Wow, that's really interesting that you could do this!"

Stern: It sounds like you had a solution, but you didn't have the problem yet?

Swider-Lyons: We kind of just wanted to do stuff and no one knew where you could go with all this. I've talked to a lot of people from the Air Force and they have a lot of their decisions from top down, "We want an airplane that can go eight hours." And we're more like, "Hey, let's see if we can go for twenty-four." It was just a technology push.

I think what we enjoy here is just trying to do stuff that's kind of nuts. The nice thing about working at NRL is that we didn't have a big acquisition plan. We're not part of the formal Navy process and the Navy protects us. They recognize that it's because we're not under these restrictions that we come up with really innovative ideas. Many of our ideas transition to the military through the private sector.

Stern: You talked before that you have this team you've put together. When you start the project, I assuming you are the project manager? How do you decide which people to pick?

Swider-Lyons: That's a great question. Well, I started off, of course, as a bench scientist. Back in the Spider-Lion days, I was on what I call the C team: me and my postdocs trying to get this together. Then someone says, "Well, that's interesting. Let's put an engineer on that." I gradually moved into what I would call the A team, and because I am the biggest advocate of the technology, I became the principle investigator, responsible for the whole team. You don't get the A team at the start. You don't get the best engineers, because they are all busy. Now, I've been able to have all the A-team engineers because everyone wants to work on my projects, because they work. But a lot of it is that I have to push it along to get these things to work. Unfortunately, I don't have time for hands on work myself anymore either.

I was in a recent dilemma about another program we're working on for an unmanned undersea vehicle, also known as a submarine. This is going to take forever to build—it's pretty big. And I said, "Let's just build something. Let's cobble something together and get it to go, because all my engineers are getting distracted with other shiny object projects." We have the shiny-object problem. The goal is to make your project the shiniest, and a lot of that comes down to getting out the PR and representing it well. But, really, it's just cool, and I'm just generally enthusiastic about this stuff. We just got the thumbs up, and should have a comparable "Spider-Lion" of the UUV built in a year.

Stern: Where does this enthusiasm come from?

Swider-Lyons: I don't know.

Stern: Do you just like learning new things?

Swider-Lyons: Yeah, I just like what I do. I've got paperwork on my desk here I've got to sign, and I've got to slog through budgets, but NRL management does a good job in providing a good environment for inventors here.

Stern: Could you give some examples?

Swider-Lyons: We have some freedom to do different things. So, for instance, I'm working on this current project, which is a huge grind. There are 40 people on it. We have engineering review after engineering review. It's a new technology, but it's a grind. I have just decided that I'm going to do a sabbatical next summer at the University of Hawaii and work on new energy systems, and everyone's like, "Okay." None of my management will stop me from going there. The UH energy institute at UH does R&D on all the alternative energy systems in the United States: solar, thermal, hydrogen, everything. So, that's the place to do alternative energy. So NRL will give me a chance to get back in touch with my energy roots, and step away from management for a while.

Stern: When you have this team, you obviously have a lot of bright people. How do you coordinate the decision-making process with all these individuals?

Swider-Lyons: It's tricky—and, to tell you the truth, it is all men on these big engineering teams. It's me and the guys. Me and seven mechanical engineers. Me and twelve mechanical engineers. First of all, I give them credit, which they are very grateful for. Unfortunately, when a principal investigator gets a big name in a field, they're often talking like, "No, I did this by myself." I'm so grateful to the pilots and to all the people who work on the project.

Stern: Do you think they appreciate the team effort after a while?

Swider-Lyons: Yes, everyone likes being on a team. But what a project manager has to do is to spend a huge amount of time communicating with everyone and making sure everyone is on the same page. Recently, for example, someone on the team brought up the fact that we didn't have the right cables for the vehicle. I mean, could you think of anything more boring than electrical connectors? So I set a meeting and then we all came to agreement on electrical connections. If you don't sort out all these details, you don't invent anything. You have to be disciplined to deal with the incredible minutiae and boring stuff.

Stern: When you're trying to come up with ideas, where do you seek inspiration?

Swider-Lyons: The lab. You basically do experiments.

Stern: But do you try to find analogies from things that are outside the lab?

Swider-Lyons: Give me an example.

Stern: Baking. There are chemical reactions in baking.

Swider-Lyons: Well, yeah. When I was doing my thesis, I couldn't figure something out, but then I realized the solution when I was baking. You just have to think about it all the time. You have to be curious and persistent about getting a perfect answer—an answer that meets all the questions.

Stern: Where there are equal signs between everything?

Swider-Lyons: Yeah, that's what I'm interested in.

Stern: So, there is a perfect solution?

Swider-Lyons: In the process of trying to get something perfect you invent something. Does that make sense?

Stern: Well, it does, but isn't there always an opportunity after you've invented something to go back and say, "I could do it even better?"

Swider-Lyons: Yeah, of course.

Stern: So are you then making it more perfect?

Swider-Lyons: Yeah.

Stern: Do ever feel that you're making mistakes or failures?

Swider-Lyons: Of course.

Stern: What's the motivation when you do an experiment and it doesn't work?

Swider-Lyons: Well, this is an interesting question. As an example, we were developing a new technology for efficiently charging batteries that we thought we could patent, but we couldn't get the experimental data together quite right. I had an intuition that it was going to work. I had a very good summer student working on the program, and he got it to work. Then I had a more senior engineer come look at it and realized that the student's results were probably due to problems with the equipment. He got it to sort of un-work. All he found was problems—they were legitimate, but we never found our way around them. I've noticed there are people who get things to un-work. Does that make sense?

Stern: What is un-working?

Swider-Lyons: Well, there are certain people who like to find problems.

Stern: Is this a glass-half-full-or-half-empty situation?

Swider-Lyons: I would say it's a different type of research. Some researchers, and I have done this often myself, just find problems. This is a valid type of research, and is important to find problems. The key to inventing though is to take the same problems, and basically turn them inside out into a solution.

Stern: You just talked about having intuition? Where does that come from? Maturity? Life experience?

Swider-Lyons: Just having a weird brain. You just have to think about stuff. For me, it's a question of hitting a space and starting to think about something and you go, "Well, that's so obvious to me." Sometimes I do a search on it and find no one has done it before, and it's so odd. But usually I find that I'm in a place where my idea is somewhat innovative. For instance, I have an idea about how to charge batteries more efficiently based on work I did for my PhD. It's something that the average electrical engineer isn't going to know because my PhD research was in this very high-end research area. So I talk to my electronics people and they go, "Well yeah, that actually might work. But I never thought of that because I don't have the background." So, for me a key part of this process is working with a whole bunch of different people and linking together different fields.

Stern: Going back to failure. How do you get over that?

Swider-Lyons: I haven't failed yet. I just keep working it.

Stern: So, they're not failures. They're just things that didn't work?

Swider-Lyons: Well, this project right now with the battery charging would be my first failure, and I refuse to give up.

Stern: What is the motivation?

Swider-Lyons: You just want to figure it out. It's just like a puzzle. Imagine you have a jigsaw puzzle and you have some pieces fitting together here that look like a bow and you have some pieces fitting together there that look like a stern and you're trying to fill in the middle. I wouldn't give up on it. I'd be so stubborn until I got the whole boat.

Stern: So, is it a personality thing?

Swider-Lyons: Yeah.

Stern: Is it something you learn in school, or someone says work harder?

Swider-Lyons: Maybe it's luck to some degree. Maybe it's having a positive experience early on. A lot of stuff is just luck. I'm very lucky that I work here at NRL and I work with a very good UAV group.

Stern: Do you have a definition for what innovation is?

Swider-Lyons: No. Don't you think a lot of success is the result of being in the right place at the right time? Like Silicon Valley. They'll tell you it's percolation therapy: people who were interested in silicon properties and computers started to get connected. But if they hadn't got connected, innovation would have dribbled away.

Stern: I believe creativity is the connection of tangential things and so you never really know what those connections are, but you have to be willing to be out there and put those things together.

Swider-Lyons: Right, exactly. You just have to slog through a lot of stuff. I hate to be trite, but it's ninety-nine percent perspiration, 1 percent inspiration. It's just a lot of work and then every once in a while you're sifting through all the sand of your job and you go, "Oh, look! There's a little diamond."

Stern: Is it just a matter of just absorbing information into your body?

Swider-Lyons: Yes, of course. Just being totally interested. Things like trash fascinate me. Why do we have all this trash? Wouldn't it be fascinating to work on bioremediation: huge industrial-scale processes, chemistry, knowing how people live, and things like that.

Stern: When you open your eyes in the morning do you just see chemistry in every facet of your life?

Swider-Lyons: No.

Stern: Do you see problems all day?

Swider-Lyons: Yes. I would say I'm a very critical person. I definitely see problems.

Stern: Are you critical of people as well?

Swider-Lyons: Yeah. But what people don't realize is I'm foremost critical of myself.

Stern: How do you deal with that all day?

Swider-Lyons: You have to be aware of it. But a lot of things in science are critical. To get a paper accepted, you criticize. The criticism is the process. It's a question of holding yourself to a high standard.

Stern: Your projects never really commercialize in a sense. The government is never going to make these things. Does that change the final product at any point?

Swider-Lyons: It changes what we patent. We patent here for a different reason than industry might. For instance, if we work on a material for the Navy, we will say we want to patent that just because if we paid for the research and that goes out to industry, we're not going pay for it again.

Stern: Do you work on multiple projects at once? And, how many?

Swider-Lyons: All the time. Ten at the moment.

Stern: Why is it important to you to work on multiple projects at once?

Swider-Lyons: It's like your eyes are bigger than your stomach. There's so much I want to do and I can't do it all myself and that's why I'm very focused on building up a team. I don't even know how to do a lot of stuff.

Stern: Do you delegate?

Swider-Lyons: I both do the work and delegate. If I'm not the inventor that's fine, too, although, not to have hubris, but my group really hasn't put out any inventions without me. One of the guys in my group has a very nice way of doing very, very careful measurements that no one has done before. But it's not patentable. I looked at what he is doing and I said, "Hey that's really interesting. Why don't we do it like this and then we can translate that to a measurement method that big Navy or an automotive company could use."

Stern: So, you're bringing finesse to it?

Swider-Lyons: I think that I have more experience. I also always try to figure out something practical from the basic research.

Stern: Along the way, have you had mentors? What do you seek in a mentor? And when do you become the mentor?

Swider-Lyons: Yes. I'm a mentor now. I mentor probably eight people informally. I don't like formal mentoring very much, but even find the informal mentoring exhausting. I had a wonderful experience in college. My professors were super-positive. They gave a lot of positive leadership and I respond very well to that. But in terms of mentors now, I haven't sought out too many mentors here. I think you try to look for colleagues who you can learn stuff from, and you always have to be learning stuff from different people.

Stern: Why do you think people now are coming to you as a mentor?

Swider-Lyons: Sometimes I inflict myself on them. I have a guy in my group getting his PhD and he just wasn't getting to a clear thesis. I had to insert myself into it. I said, "You have got to find a question to answer." And I gave him three questions to answer, and I said, "Pick one. I don't care." He was just wandering. But it was funny because I had been wandering in graduate school without much help from my advisor, and then, Fernando Garzon, who had been my undergraduate research advisor when he had been a graduate student, was off at another job but decided to intervene with my graduate research, and basically got me to answer a question. It really saved my graduate career. So hopefully I passed on some of the mentoring I had had before.

What you can't teach is that you have to work hard at anything to be good. There is no way to be successful without working hard. But you have to work in a certain direction and be disciplined. You have to open your mind in different ways.

Stern: Does being in a military environment broaden your horizon or close it in at all?

Swider-Lyons: Well, both. When you work for the military, you have a sense of being part of a very large organization.

Stern: I meant more the strict discipline and chain of command and following orders.

Swider-Lyons: Oh, not so much here. Here it's very loosey-goosey in a sense. They recognize that creative people don't need to be browbeaten. We do have a chain of command and obviously security rules that you don't have with a university. I set up a system in my group and said, "Just follow it." And when people complain, I'm like, "Fine, don't work for the military. I don't write the rules." You just learn not to fight the rules. I think that comes with maturity as well.

Stern: You said before there are creative people here. Do you have a definition of what a creative person is, as compared to a non-creative person?

Swider-Lyons: You know there are intuitive people.

Stern: Is creativity just this inbred intuition?

Swider-Lyons: I think it's personality-based. You might be able to learn it. I don't know. What do you think?

Stern: I think it's inbred in you. You can get better at it, but if you're not sensitive to certain things—if you're not inquisitive—there's no way to teach it.

Swider-Lyons: And you need a certain amount of confidence short of hubris. I guess in that respect it's personality-based, and that is somewhat just what you're born with.

Stern: How do you find new projects? Do you have the liberty to develop your own problems?

Swider-Lyons: Yes, that's what we do here. I have to pitch new ideas every year or two to get research funding. They don't just shower money on us and we just do things. We have to write a proposal and have a schedule and meet milestones.

Stern: Have your projects gotten out into the marketplace with actual usage?

Swider-Lyons: Not my Navy ideas yet. We're working on it. Although I did work for a year at Johnson and Johnson, and was on a patent there, which I believe made it into the consumer market for children's chewable Motrin.

Stern: Do you think that anything will change once it gets out there—once you have, in a sense, consumer feedback from soldiers in the field?

Swider-Lyons: Yeah, if hydrogen fuel cells go off, I may not even know what happens.

Stern: When you say "go off" you mean?

Swider-Lyons: If they're deployed. I don't know what they will be doing. The issue with the military is that they often they want things which, because of budget constraints, you can't offer them. It's a very complicated process to get something into the military. The chief of naval operations sent down a query two weeks ago saying. "Why hasn't this transitioned?" Well, there's no requirement. Someone in an office has to basically develop a requirement saying, "We need a thirty-five-pound UAV that can fly for a day." Even though there is a military want, there has to be a formal requirement process. And that takes a while. In some circumstances, people have done better by going out through the private sector and having the military pick it up later when it's a tangible object that they can see and has a SKU number on it.

Stern: Any advice you would give to be an inventive person?

Swider-Lyons: Put yourself in an environment where you're going to succeed. You have to be patient, and don't be a prima donna. You have to work within the system. You have to respect everyone one in your chain who is helping you—from the administrative people to the budget people. I don't like people who come in here and say, "Why do we have to do a blah-blah review?" You know what? You have a boss and your boss has a boss and your boss's boss has a boss. Everyone has a boss and everyone is responsible to someone, and what I don't like is when some scientists think, "Oh, well, I'm doing science—like I'm an artist. I should just be able to do that."

In my case, I am working for the US government, and there are rules in place that I need to follow. You can be very inventive within the rules, but you have to be respectful of all people who are helping you make this environment for you. You have to be grateful that you're in a position where you can invent something and accept the downsides. You have to slog through paperwork, the usual thing of any job. The major advantage that we still have here at NRL is that we can have multi-year development cycles. As I said before, I can work on things for years, sometimes only a little bit, but it helps to be in an environment that has some patience.

Stern: You mentioned patience as an attribute. Why is that important?

Swider-Lyons: Sometimes it's this puzzle and you're missing a puzzle piece. Sometimes you might have to shelve it for a while until that puzzle piece becomes apparent. It doesn't always come just like *that*. [snaps fingers] Often you have to circle back a year or two later and take another look as a new technology comes out. You're relying on other things to come from other people. You have to patient and keep trolling for new stuff.

Stern: This technology you're developing now. Do you ever think where will it go in twenty years or thirty years?

Swider-Lyons: If the stuff we're working on with the unmanned air vehicles and the unmanned submarines works, it will revolutionize the Navy.

Stern: Could this technology be applied to cars?

Swider-Lyons: It is car technology. We are adapting it to unmanned systems.

Stern: Could it go into the consumer marketplace?

Swider-Lyons: Yes. This is a type of thing where the consumer market has been pushing it, but it's not getting picked up fast enough. The military can adapt it faster than the consumer market.

Stern: Why do you think the consumer market is not picking it up?

Swider-Lyons: Because, after more than 100 years of development, gas engines are cheap. Hydrogen fuel cells are expensive for the consumer market, but they're cheap for the military because they enable new missions. I cannot understate but if we can do long-endurance unmanned vehicles for the military, they will revolutionize the Navy twenty to thirty years from now. They will change how they do business.

Potentially you could send out vastly cheaper unmanned systems to do the job instead. The Navy could have a persistent fleet of unmanned air and undersea vehicles out all the time. They're cheap and relatively small. It's like how you are using your iPad now instead of your desktop computer, a recording device, and digital camera. You still need the big equipment for high quality work, but you can get a lot of work done with just your iPad.

What's fascinating right now with energy technology is that so much of the future of energy technology is the controls and autonomy. One of the things that I'm really looking forward to in my life is when cars become autonomous, eliminating the risk and waste in how people drive—stepping on the accelerator, swerving, and texting. You live in Oregon, right?

Stern: Portland.

Swider-Lyons: Not too bad.

Stern: We ride bikes all day.

Swider-Lyons: Yeah, here it's different. I drive on Interstate 395 to get to work. And you don't have the weather we have here either. Either snowing or 100 degrees is not good biking weather. But if you have people in autonomous cars? If you have a satellite or maybe a series of UAVs for a secure uplink, then you could literally just hand over your car and then it would drive. There are sensors on the car so you don't get in accidents and you don't get in traffic jams. You don't have sixteen-year-olds dying while texting, and it helps older people, too, because they could actually have a lot more freedom. I don't know how to do autonomy, personally, but lots of people are working on it all over this place.

Stern: You see your technology as being part of that?

Swider-Lyons: Yes. So, this whole movement for long-endurance electric systems that my work is part of is going to enable autonomy, and autonomy is going to free us from drudgery and danger and waste. Thirty years from now you will not be driving your car. It is going to be huge energy savings because the controls will drive you the most efficiently. We're doing pieces of this technology for the military. I think it's going to just revolutionize how people live.

Stern: Any advice that you'd want to give to an inventive person?

Swider-Lyons: Oh, I do have some advice for science. My father was a theoretical physicist. I struggled with quantum mechanics in college. He said, "Don't worry about it. You don't need it." "I don't?" "Yeah." So, I think if you want to do science or technology, don't get intimidated by what you don't know, because you can find people to help you. A friend of mine said recently she dropped out of chemistry because she didn't understand quantum mechanics. I'm like, "Well, I understand it a little bit, but I am still a professional scientist."

Stern: There's someone you can go ask to solve a quantum mechanical problem if it ever comes up?

Swider-Lyons: Yeah, the guy next door here does it. There are geniuses upstairs from me. These guys just sit around developing the code for quantum-computing things. They are right upstairs. I don't have to do it. I think that's probably my piece of advice: If you like to do stuff, keep doing stuff, and just ask for help. In the worst case, you have to put another person on your invention. Don't be a sole inventor. That's stupid.

Stern: Outside of the work you're doing, do you have a favorite product or technology out there—or non-technology?

Swider-Lyons: I love my iPhone.

Stern: Everyone loves their iPhone.

Swider-Lyons: I know, but look at this thing! It is so awesome. I've worked on this portable power stuff that is big and clunky and I'm like, "Wow, I can hold this in my hand and it feels great."

Stern: When you're at a dinner party, what do you tell people that you do?

Swider-Lyons: I tell them I'm a scientist. Yesterday I was in a pool with a girlfriend of mine, and I told her I sold a program to build a submarine. She's an accountant in a law firm and she's like, "Yeah, whatever." I'm just so excited we get to build this little sub because there's a big submarine next. I want to build a little one first.

Stern: How big is little?

Swider-Lyon: Twenty-five feet. I'm like, "Whoopee, we get to build something!" And all the guys are like, "Whoopee, we get to build something!" It's just exciting to build something. Who gets to come to work and build stuff? It's fun. I can't imagine what it's like to just be doing accounting for thirty years.

Stern: Do you plan on retiring, and what would you do?

Swider-Lyons: Good question. I'm forty-six, so I have a way to go. That's actually a very good question, because right now I'm thinking, "I could retire so I could sleep." I talk to people here, and when they hit thirty years with the government they say, "Now that I can retire, it's not so bad, and I don't know what else I'd do." I guess you need to know how long you're going live, right? What are you going to do? But this is a dilemma. In my neighborhood everyone has perfect lawns and their houses are all clean. I don't want to die and be known as the person with the clean house. I've never seen anyone's obituary that said, "They had a very clean house." I mean, you don't want rats in your house. My house certainly isn't messy or anything. I love my Roomba. Oh my God, that's an amazing autonomous invention. One of the best uses ever of DARPA funding. Yes, I love my Roomba.

Stern: That was invented by Helen Greiner. She's Chapter 7.

Don Keck

Corning Incorporated (Retired)

Dr. Donald B. Keck retired in 2002 as a vice president and research director for Corning Incorporated. As part of Corning's Optical WaveGuide Project team along with Robert Maurer and Peter Schultz, he invented low-loss optical fiber in 1970. This accomplishment launched the optical fiber telecommunications revolution and enabled the Internet. More than 1.6 billion kilometers of optical fiber are deployed worldwide today. Dr. Keck's research areas include molecular spectroscopy, gradient index and aspheric optics, guided wave optics, optical fiber sensing, and optical fiber waveguides and communication. He is the author of more than 150 papers and he holds 36 patents.

From 2002 to 2004, Dr. Keck served as CTO for the Infotonics Technology Center. Currently, he is vice-chair of the Invent Now board of directors and serves on the committee that recommends National Medal of Technology and Innovation nominees to the President of the United States. He is a past member of the Congressional oversight board for the National Institute of Standards and Technology (NIST), former board chairman of the Optoelectronics Industry Development Association (OIDA), and past president of the National Inventor's Hall of Fame. He served on the boards of directors of PCO, Inc. (a joint venture of Corning, Inc. and IBM), and the Optical Society of America.

Dr. Keck is an inductee of the National Inventors Hall of Fame and a member of the National Academy of Engineering. He has served on several U.S. National Research Council panels. He is a Fellow of the Optical Society of America and the IEEE, and an honorary member of the World Innovation Foundation. Other awards are the Distinction in Photonics Award, the Department of Commerce American Innovator Award, and the U.S. President's National Medal of Technology and Innovation.

Dr. Keck received his physics degrees from Michigan State University. He is a distinguished alumnus and serves on the College of Natural Science Advisory Board. He received an honorary doctor of science degree from Rensselaer Polytechnic Institute.

Brett Stern: What is your background? Where were you born? Tell me about your education and your field of study.

Don Keck: I was born in the Midwest—East Lansing, Michigan. My father had a doctorate in physics. My grandfather lived next door and was a jack-of-all-trades. I had the benefit of a father that knew the science and a grandfather that had a wealth of experience. Both my father and my grandfather were of the mind that if you wanted something, you made it. You didn't buy it. They were always creating things that did something they wanted to do around the house or around the property.

I grew up a quarter mile from the Michigan State University campus. The college was always there in my life, and it was perfectly natural that I would at least do undergraduate work there. All three of my degrees in physics are from Michigan State University.

Stern: Growing up, were you following in your father and grandfather's footsteps—inventing and improvising things?

Keck: Absolutely. You didn't have LEGOs and all that when I was growing up. I made my own toys out of cardboard boxes. I got interested in electronics and made a tachometer for my boat. There were always equipment and tools lying around, and I enjoyed and profited from learning to use them—as did Corning later on.

Stern: Could you give a brief background in layperson's terms of the invention of the low-loss optical fiber material that you developed?

Keck: I graduated from Michigan State in 1968. It was a wonderful time. There were all sorts of jobs available at that juncture. I had seven or eight job offers all over the country. I picked the one at Corning offered to me by physicist Bob Maurer. Found it interesting. Bob described a need for doing better optical fibers. Fiber optics had been around since the early 1950s, but optical fibers could only transmit light a few feet.

The laser had been invented in 1960, as you probably know. Although few people thought the laser held much promise for optical communications displacing the coaxial cable civilian communications infrastructure. However, the British military was interested in the laser for military applications—transmitting high-frequency radar signals over long distances. The coaxial cables of the day had fundamental limits: all you could do to get longer-distance transmission on coaxial cables was to make them bigger and bigger. All the telecom firms in the States and developed countries were taking that "engineering approach."

Back in those days, researchers tended to stick pretty close to their home laboratories. But Corning sent out a technical liaison, first William Shaver and later Gale Smith, to tour laboratories around the world. When Shaver came back from one of these tours in June 1966, he wrote a memo to Corning management reporting several British military groups were interested in fiber optic research aimed at long distance light signal transmission. Later we found other laboratories were interested in the same topic and doing their own research.

Corning management asked Bob what he thought of the idea and if he thought it was possible to bring attenuation below twenty decibels per kilometer. Bob was always interested in pushing the limits of science, and replied that he thought it was possible. He soon picked out a glass candidate. He had an experience base built on fundamental physics on how pure you could make glass. Ultimately, the transmission of light in a transparent material is governed by two factors: how much the light is scattered out in the material, and how much of the light gets absorbed in the material and turned into heat. Bob had extensively studied the light scattering mechanism, which was the ultimate fundamental transparency limit. He had a catalogue of all the glasses in Corning's repertoire that might be candidates for a really transparent glass. Fused silica was the best material that he had measured.

Bob wrote a memo suggesting that fused silica be used, and he and a summer student went ahead and tried to make use of it for fiber. They succeeded in making a fiber, but the loss was still very high, and they couldn't transmit light through more than a few feet of the material. But Bob was fundamentally intrigued by his light scattering measurements, which were very low. He made the request to his management to add people to this project.

Authorization rattled back down through management to Bob from the top—Bill Armistead, Corning's VP of Technology. A young materials scientist, Peter Schultz, who had just been hired in the chemistry department, was lent to Bob to dope his fused silica. The three of us—Robert Maurer, Peter Schultz, and I—were put together to work on the project. Bob was manager of the department at the time, so he couldn't spend full time. Pete and I were really the two kids who "didn't know it couldn't be done."

Stern: Schultz's background was chemistry. Yours was optical physics. Can you talk about collaborating with someone with a different background?

Keck: I knew how to study materials by measuring their spectral transmission and thereby studying their basic optical properties. Those, plus ultimately making the fiber, were my essential duties. The interaction between Pete and myself was essentially the "collision of disparate minds". We were converging on the goal of making this ultra-pure material into a fiber from two different directions. Pete came from the materials standpoint, focused on making new doped fused silica materials and studying their properties. I came from the physics metrology standpoint, and focused on understanding the fundamentals of light transmission in optical fibers.

Stern: In doing all this research, did you limit yourselves to solving the theoretical breakthrough problem set by the British, or were you guided by the potential applications for this technology?

Keck: Once the British had put this thought in our minds, we began looking around at other laboratories that might be interested in something like this. Along about 1969, we entered into a collaborative relationship with Bell Labs. It came about as a result of a cross-license agreement that Corning needed for the transistor. At the time, the transistor was coming into

its own, Corning had entered into that business area through the acquisition of Signetics and we needed transistor patents.

We entered into a cross license with Bell Labs, and one of the technology items we put into the cross license had to do with fiber optics. Scientists on both sides got together to explore where we might collaborate in this list of technologies in the cross-license agreement, and fiber optics happened to be one of the ones that was picked. Bob and his counterpart, Dr. Stewart Miller, at Bell Labs began an interaction. We got our scientists together. They would tell us what they needed in the way of telecommunication. We would tell them what we were doing in fiber optics.

Stern: Would you say people appreciated the potential at the time?

Keck: There was great skepticism. Most senior scientists were naysayers who said you'd never be able to create a glass transparent enough to use fiber optics in telecommunication.

Stern: Did you have the same skepticism?

Keck: No, Pete and I were just curious young scientists who didn't know it couldn't be done. Bob had done calculations that indicated it should be possible, and the next question was whether we could physically pull off a process to do that with the right materials. I was in the Corning group that went down to Bell Labs to find out their interest in it.

Stern: Would you say the naïveté is an important prerequisite for an inventor?

Keck: The inventor has to be optimistic at all points. First, you have to believe seriously it can be done, and then you have apply your inventive skills to solving the problems that you uncover as you pursue that end.

From 1968 through 1970, I made better and better measurements. I started out just using a single laser to measure the loss at one wavelength. Later on, I built a spectrometer to measure the spectrum of the light transmitted through our fibers to more precisely determine the sources of loss. I came up with the markers that told us what problems we had to solve. Initially, absorption was the problem. We figured out how to solve that with heat treatment of the fibers.

Then scattering was the problem. We found that the light scattering was essentially coming from the "rod-and-tube" process for making fibers. This old-school process involved sticking a solid glass rod inside a glass tube and heating them both in a two thousand–degree centigrade furnace so that the glass essentially fused together and started running out the furnace end like taffy. The glass picked up electrostatic charge, and all sorts of debris would stick to the rod and the inside of the tube, producing light-scattering centers that radiated all of the light before it ever got to the end of our fibers.

Another factor we knew from the physics was that the fibers that we were after were a special kind of fiber called "single-mode." They had to have a very small core to carry light in what's called a single optical mode—thereby giving the highest possible bandwidth or information carrying capacity. But the fibers made by the rod-and-tube had a very large-diameter core and a very thin cladding glass around it. We wanted just the inverse for telecommunication. Basically, the fiber itself is about the size of a hair, and we needed the core of the fiber to be a tenth of the diameter of a hair for the telecommunication application.

Stern: How did your group brainstorm possible solutions?

Keck: We sat around the table and tossed out ideas. We made a list of them and tried to rank them in some fashion. In one of these meetings, I suggested, "Why don't we deposit a thin film on the inside of the tube."

Stern: Where did that idea come from?

Keck: I picked up that idea on one of my interview trips when I was looking for a job. These ideas come from the wealth of the experience base that you pick up, and they can come from the strangest places. I think the creative spirit in human beings instinctively files these things away. The inventor, the clever person, resurrects things from their experience base that they apply to a new problem. Films for transistors and semiconductors had been invented already. They were applied to encapsulate semiconductor products to shield them from the atmosphere.

Stern: After your brainstorming session, what was the next step to prove or disprove the concepts? Did you sketch? Prototype? Work in 3D? How did you go through the possibilities?

Keck: There is any number of better ways of doing it with the computer modeling of today, but we didn't have that technology. Your fundamental process is identifying all the things that you possibly can.

As a young scientist, I was learning who all the people in the lab were, so every time I introduced myself, I'd toss out the problem that was confronting us. I picked up other ideas, and then took it up to the team to prioritize the best approach in terms of resources. I worked on methods for depositing the film for four months. I succeeded in depositing the film, but it was pretty obvious the approach I was using wasn't making it.

Pete's process involved spewing fine particles of the glass out of a burner. We hit on the idea of trying to squirt some of these particles from his flame through a small hole in the center of the tube. At first, the material wouldn't go in the small hole. The burner was about an inch in diameter and the hole was a quarter of an inch and we just weren't getting much stuff in our tube. There was a vacuum cleaner in the corner of the lab, and one of us grabbed that and stuck it up into the end of the tube—and low and behold—it sucked the doped fused silica glass particles that we wanted for the core glass through the tube, depositing a nice thin layer. Today that process is better known as "inside vapor deposition" [IVD].

We heated our IVD preform in the high temperature furnace, and we drew our fiber out of that. On the second or third time we tried to make a fiber using that process, we made the low-loss fiber that essentially changed the world.

Stern: Concerning the lab failures, what is the motivation to keep going forward and trying something else?

Keck: Curiosity and having a goal. Defining a problem, a large portion of it is to set the right measure for success. In our case, it was the transparency of the glass. We didn't have to worry about the number of bits you can send through the fiber per second. Bob kept a chart of the improvements that we were making in the transparency of the glass, which he would show upper management in those two years that we were working. That gives you incentive for continuing the project.

Stern: Why do you think Corning gave you the liberty to just keep going down this road?

Keck: We were able to demonstrate progress to upper management, but, quite candidly, I'm not sure how much longer they would have continued to support us. Years later, we found that Bill Armistead had a list of the top twenty projects in the lab—and we were number seventeen on his list. So, how long before some other possible application of glass that needed resources would have knocked us off the list we'll never know?

Stern: At what point did you come up with something that had commercialization appeal?

Keck: It was a long, long time before we knew if it were going to revolutionize the world. Even after we had the breakthrough that said twenty decibels per kilometer—that is we could transmit one percent of the light through one kilometer of our fiber—it took twelve years before we knew we had a business. At that time, upper management was into glass for consumer applications: dishes or glass for television. We knew nothing about telecommunication. It was the interaction with Bell Labs that ultimately proved to be helpful for us.

We took a sample down to Bell Labs and they corroborated our measurements. They came back to Corning upper management and said, "Do you know what you have here"? Upper management was getting input since 1970 on how important this could be to the telecommunication industry. Although they had no certainty that Corning would get business out of it, they were supportive at the top levels of the corporation. Of course, at the lower levels, our group was interacting with Bell Labs, and they were encouraging.

Stern: During this twelve-year process, was this the only project you were working on then?

Keck: Yes. I'd say by 1972 on, we were pretty much convinced that it could find its way into the marketplace. There was just the question of proving to the potential customers—the Bell Labs and the telecommunication firms around the world—that in fact the fiber could do everything they needed it to do.

I had set up lasers that could measure the bandwidth of the fiber and the number of digital bits per second it could send down the fiber. We then designed fibers that would do more and more. People had to be able to handle the fiber, so Corning went to all of the electrical cable manufacturers and tried to get them to venture with us to produce fiber cables. Most of them said it would never go, so they backed out.

In the end, Corning formed its own optical cable company with Siemens, called Siecor. We learned how to cable and we made the first optical cables, which Corning demonstrated under a contract to the Navy in 1975. This was five years after the invention—we finally had a practical cable. There were no ways of connecting the fibers. So, we had to invent connectors and show them to potential customers, stating that we could splice these things together. So we invented the process for splicing fibers, aligning cores that were about a tenth the diameter of a hair.

Stern: Was the group expanding, or was it still you and Pete?

Keck: No, it very definitely expanded. By 1975, we were organized into managerial groups, groups of five to six scientists and an equal number of technicians. By '75, I was running one. Pete Schultz was running one. It was a sizable project, considering that we still didn't know if we had a business. In 1976 Corning built a pilot plant to get samples out to potential customers.

One of the wonderful things that one of the marketing people, Chuck Lucy, did around 1971 was to set up some joint ventures with five telecommunication firms in Europe and Japan. For several years, every six months, we traveled around and did technology exchanges with them. They taught us about telecommunication and we taught them about glass fibers. They said there were worries that "glass breaks." So Bob led a project to demonstrate the strength of glass. It was a two- or three-year project that demonstrated that glass fiber is in fact stronger than a steel wire the same size and that it could withstand the stresses to which it would be subjected. We gave them samples of fibers that they could run experiments and field tests on. In 1976, we had a pilot plant that could produce roughly one thousand kilometers of fiber a year. Still no sign of a real business. People were testing the fibers and running experiments.

Stern: What do you think was the apprehension in applying this technology?

Keck: All along the way you had to keep knocking off problems that they showed to you. For example, hydrogen from some of the cable materials would diffuse into the fiber and begin to increase the absorption. It took a couple years of research to solve that potential stopper. You have to be pretty careful about how fast you move on putting in a new infrastructure to handle telecommunications in your nation.

Stern: Along the way, would you say you had any mentors? And, if so, what did they offer you?

Keck: I wouldn't say "mentors." But what I found happening was that in these interactions company to company, I'd built up good personal friendships that were very helpful in eventually moving the technology forward. We could candidly share with the people with whom we had the joint ventures. We shared what was going well and what was not going well, and we eventually solved whatever problems were on their minds at that particular time.

I haven't mentioned so far something that I find most intriguing. The four things that were really important in creating the telecommunication age and the Internet were all invented within about six months of one another.

Stern: What are those?

Keck: In 1970, we invented the fiber. Then the room-temperature laser was demonstrated. The first Internet experiment, called ARPANET, was run at UCLA. Ted Hoff invented the microprocessor, or PC chip, in early 1971. Before these four advances, all you had was telephone. As soon as you began transmitting the data, then everything broke loose.

Stern: Do you have a definition for invention and innovation? Is there a difference between the two?

Keck: The innovation includes everything that's necessary to get the thing into some commercial application. The invention is really the instantaneous creation of something new and unique. You could have many inventions that make up eventual innovation. By the time that the fibers began to really get commercially employed, Corning had some four hundred inventions—all sorts of things from the glass materials, to materials and processes for coating the fiber, to the cable and the connectors—that underlay the innovation of getting the fiber optic technology into widespread telecommunications application.

Stern: Along the way, would you say that you have found certain skill sets that are required to be an inventive person?

Keck: I have alluded to this earlier in our conversation. To gain the experience base that supplies the depth of knowledge to solve whatever problem might come along, the inventor needs curiosity and inquisitiveness: What's going on? Why does this work?

Stern: You also alluded to a then-and-now contrast. In those days, you were very hands-on in making your inventions work. Today, everything you did then could be computer-modeled three-dimensionally. Are there advantages for one way or the other?

Keck: I guess my thought is, "Whatever works." Computer modeling has gotten better and better over the years. When we were doing our work then, computers really weren't around. They were in the university. The math statistics group at Corning had a big IBM mainframe computer that could tackle really difficult problems, but modeling capabilities just didn't exist.

I had grown up with computers that were basically doing the statistical modeling of molecular spectrum. So, I was reasonably familiar with doing this and eventually got the first computer in the lab. I was actually taking data off the optical bench that was in my lab directly into a computer. Would I have used 3D modeling if the capability had existed then? Sure, you use whatever tool is available to you. Modeling can circumvent a lot of dead ends. But, at the end of the day, if the thing has got to go into a customer's hands, you have to make something tangible.

Stern: Do you have any advice that you would offer to an inventor?

Keck: A good idea will stand the test of time. Time and time again, I found that I had dredged something up out of a past experience and applied it to a particular problem. Corning can point to many examples of material that didn't fit the particular market at the time of its original creation, but was later uplifted and retrofitted to unforeseen applications.

I think inventors need to hang onto that wealth of experience, ideas, people, and technology that they have built up. What I found working on problems over the course of my career is that you have got to contact as many different people as you possibly can to pick their brains. Don't operate as though you've got the only ideas in the world. You may have some good ones, but if you listen well to somebody else's thoughts on a particular problem, you might come away with a new idea or a new direction.

And, of course, there is no substitute for having a good base of knowledge—especially a deep knowledge of physics or chemistry—and a lifelong experience base of having done a lot of things and being curious about how things work. When I was young, I loved to tear things apart to figure out how they worked.

Stern: Do you consider yourself retired now?

Keck: Oh, yes. I stay as busy as I want to be. I've been working a little bit with a friend out in California who is starting a new company. And I'm on a couple of boards. I like puttering around in my workshop. I am the substitute organist at our church. We even play some golf and I'm still reasonably adept with a slalom water ski. Add to that some volunteer work and interacting with the grandkids—life is good!

Bob Loce

Principal Scientist
Xerox

Robert Loce *is a principal scientist and technical manager in the Xerox Research Center Webster in Rochester, New York. His current research activities involve leading an organization and projects into new video technologies relevant to transportation and health care. He has 157 patents and numerous publications in the areas of digital image processing, image enhancement, imaging systems, optics, and halftoning.*

Loce is a Fellow of SPIE—an international society for optics and photonics, and a senior member of IEEE. His publications include a book on enhancement and restoration of digital documents, and book chapters on digital halftoning and digital document processing. He is currently an associate editor for the Journal of Electronic Imaging. He has served as an associate editor for Real-Time Imaging and IEEE Transactions on Image Processing.

A native of Rochester, Loce joined Xerox in 1981 with an associate's degree in optical engineering technology from Monroe Community College. While working in optical and imaging technology and research departments at Xerox, he received a BS in photographic science (Rochester Institute of Technology (RIT), 1985), an MS in optical engineering (University of Rochester, 1987), and a PhD in imaging science (RIT, 1993). A significant portion of his earlier career was devoted to the development of image processing methods for color electronic printing.

Brett Stern: Can you give me some background? Where were you born? Tell me about your education and your field of study.

Robert Loce: I was born and educated in Rochester, the same city I live in now. As for my field of education, I switched majors a lot, which I think fostered my ability to invent. In community college, I first focused on chemistry, then switched to geology and earth science before completing my associate degree in optics. Then at Rochester Institute of Technology, I earned a four-year degree in what was called "photographic science" at the time. Then I completed a master's in optics at the University of Rochester. At RIT again, I earned a PhD in image processing and digital imaging, which happened to be the first doctorate in the world awarded in the field of imaging science. Nine years later, I passed the US patent bar exam and became a registered patent agent. So you might say I've jumped around quite a lot in my educational career.

Stern: What were you like as a kid? Would you invent or build things?

Loce: I don't know that I invented or built much as a kid. I took a lot of things apart. I liked science. The Apollo missions inspired my generation. Even commercials on television had slogans such as "better living through chemistry." Science was very inspirational when I was young. I'm concerned that the current generation doesn't have that same motivation and isn't going to be as inventive in some ways.

Stern: Could you tell me about the field that you're in right now and the types of things that you've invented in your career?

Loce: I have one hundred fifty-seven patents in the realm of optics and imaging. They range all the way back to copier optics, through optics for digital printers and all types of image processing that went into digital printing. Most of my work the last few years has been in video processing in the areas of health care, education, and transportation imaging.

Stern: In your career, have you had the opportunity to pick and choose the projects you work on? How do the problems get put in front of you?

Loce: When you work for a company, you have to be bounded by the parameters that the company's business is in. But I've pretty much been left on my own. Managers have pretty much let me run because they know I will stay within reasonable bounds. They say, "Sure, go ahead and invent it."

Inventing at Xerox occurs within the prevailing domain of corporate focus. For years, Xerox was primarily committed to the development of printer technology. But a few years ago, we switched our focus from printer technology to business services. "Business services" can mean many different things. It can mean helping healthcare providers and insurers manage information, management of smart transportation networks, and aiding in processing information in the education system. Stating it simply, we changed from operating primarily in one field with its technology needs to many fields with more diverse needs. We researchers were given a great deal of leeway to invent in these new service areas and our managers challenged us to show what we could do. We assembled multiple teams, and launched projects that set off in several different directions.

Stern: Do you go into the marketplace to find problems?

Loce: Yes. There are many different ways to find problems, from the systematic to the accidental. Sometimes you find your problems just by observing everyday life. For example, when I go to the doctor and have trouble getting the old X-rays of my hips, I wonder if I could make that system more efficient. Or if I'm not comfortable with how my kids' schools teach and grade assignments, and how results are communicated to the students and parents, I wonder how I would do that differently. The dramatic broadening of our research space at Xerox allows us to draw on problems we encounter in daily life, and fashion ways to improve them.

Stern: Do you put together some type of problem statement or design brief? What are the steps that you go through to start and finish a project?

Loce: Often you try to visualize the solution in your head as an operation in outline form. You see the problem in front of you. You visualize the way you'd like it to operate. You start thinking of the steps in an invention to make it happen. And finally, you set about proving that those steps or elements actually do solve the problem.

Stern: When you're working on these projects, are you working by yourself or are you part of a team or a group?

Loce: Very much part of a team. I work in a very inventive group. Many of the people that I work with get five to ten or more patents a year. I think to patent you have to be presumptuous enough to think that you can always do it better. To state it more positively, the people I work with are comfortable hazarding ideas that could be wrong and trusting the discussion to improve them. I've worked in other settings where if you say something and it's not quite right, people look at you and just say, "Well, that's wrong and it's not going to work." End of discussion. But my group here in the research lab collaborates to improve ideas rather than shoot them down.

Stern: Are there multiple disciplines in this group?

Loce: In this group, there are a lot of electrical engineers, computer scientists, and optical engineers, some statisticians, a physicist, and a medical physicist.

Stern: Do you work on multiple projects simultaneously?

Loce: Usually a good number of us are working on multiple projects. Some people like working on one, and they do a very good job on one project. Some of us are a little bit more scatterbrained and could be working on three, or four, or more at once. I tend to be working on three, four, or five different things at once.

Stern: Do you find that working on multiple projects gives you the flexibility to alternate intensity with stepping back to ponder some of the directions you may be going into?

Loce: That's part of it. In addition, a fruitful way to stimulate inventions in a given field is to see how another field solves analogous problems. How could I translate their approach, or even the steps of their solution, to my very dif-ferent setting? There's a maxim among optical engineers that if you really want to understand optics, work in infrared. Working in a different part of the electromagnetic spectrum helps you understand your own problem better. There's a similar saying in patents: "Only God creates from nothing. Man must create from known elements." A lot of invention is putting known things together in new ways.

Stern: Do you have a preferred definition of invention?

Loce: Not really. I hear people debating what invention and innovation are. I tell them, "You go ahead and debate, and I'll go ahead and invent useful things."

Stern: Is there a difference between invention and innovation?

Loce: Innovation is successfully bringing new things to market or society, whereas invention is the Aha! step that actually enables the innovation to work.

Stern: Does an inventor need particular skill sets that are different from those of the average person?

Loce: I don't know about skill sets. I think it helps to have exposure to multiple disciplines. My early studies in chemistry and geology, for example, proved to be useful in my later work in optics, image processing, and video processing. I think coming from a varied background, you can see a problem from differ-ent perspectives. Each one of those fields tends to view the world a little bit differently—so when a problem comes in, you can converge on it from several directions. The most valuable skill, I think, is more a temperamental disposition than a specific technical skill. It's the willingness to attack and solve a problem by chipping away at it with incremental improvements and considering varying definitions of the problem and different points of view.

Stern: So inventiveness is almost a personality trait. Do you use any particular method, or seek inspiration from any particular place?

Loce: No, no particular place, and there's no particular inventive method or process. Different people invent in different ways, and the same person invents in different ways at different times. I like hearing other people's screwy ideas. It's sort of like you don't have to be the best at a sport to enjoy participating in the sport and watching other people do it. I don't have to be the originator of every single invention to enjoy being around new ideas. People bring their half-baked ideas to me, and I have some half-baked ideas of my own that I bring to them. And between us, we move forward and finish baking them.

Stern: Do you need to be in your office working nine to five? Or does the inventive process happen less rigidly?

Loce: As much of the process happens outside the lab as happens inside it. For most of us working in research positions here, our workday is intertwined with our home life. You wake up in the morning and jump on the computer. You're taking a shower, you're thinking of this. You're at your kid's baseball game, you're thinking of that. You're hiking in the mountains, you're thinking of something else. By the same token, while you're at work, you might be thinking about hiking in the mountains. The two flow together.

Our work life isn't so rigid that you are supposed to be here nine to five, so there is room for natural flexibility and flow. For example, I just came back to my office here from a four-mile run, so those breaks really help. If you sit here all day chained to a keyboard, there are physical and psychological reasons that you're not going to be as productive.

Stern: So the physical output balances the mental output?

Loce: I think so. Research proves the role of exercise in keeping your brain vital. I do all sorts of cardio exercise. I have advanced arthritis in one hip, but I just went out and ran four miles. Obsessive personalities such as mine help in maintaining these pursuits. I'd like to think that my obsessions are synergistic. When I go out and run like crazy, I come back, and I feel that my body and brain are all oxygenated and ready to work.

Stern: Do you use any particular tools to ideate or brainstorm? Do you sketch? Do you prototype? Do you three-dimensional model? Or do you start with the back of a napkin to get a final idea?

Loce: I don't know if I've ever really written on a napkin, but I like having marker boards nearby. Sometimes I'll write on a piece of paper, but three-quarters of the time I'll write on a marker board. It's easy for me to quickly change and sketch something: I can sketch more or I can just wipe it away, and so on. Now a lot of ideas start out on a marker board. You are sometimes alone doing that, but sometimes with one or two other people, there with your erasers and your pens, and you just keep hacking away until you get something you are all comfortable with.

Stern: Are you sketching formulas or three-dimensional designs of objects?

Loce: Usually both. Not always both, but often it starts with a visualization of the physical layout—let's say, a camera system on a highway. Well, this camera here needs to be calibrated in a certain way and it's going to have this timing with these other cameras. I'm sketching roadways and cars on roadways and trying to figure out certain issues with flow, and I'm writing formulas while I'm sketching cars.

Stern: Do you get to the point where you are actually prototyping in 3D, where you go into a shop and start soldering, hot-gluing, and duct-taping stuff together?

Loce: I don't personally do that as much as I used to, but my teams are doing that constantly with just about every invention that we come up with. Right now, I am working on new types of camera systems. We are putting them in all sorts of unusual places and observing all sorts of things, and proving out our inventions before we try to patent them. I'm involved in helping to design the experiments, but I don't actually perform them anymore.

I came here to Xerox as a technician with a two-year degree, so I spent years in the optics lab, tuning optical instruments in the dark with my flashlight tucked into my armpit. Back then, I had to physically make everything, and go to the model shop and cut things up. I gradually migrated from that to a more theoretical approach. Hands-on experience gives insights that theoretical work alone often cannot provide. Besides learning from experimental data, being a part of the experiment helps you visualize the problem and your system. You are better able to see alternatives in your mind's eye.

Stern: Any career in invention has its share of big failures and mistakes. Without being specific, could you talk about what you learned from failures and how they gave you insight to go forward?

Loce: How about I give you a little twist on that. I once had a manager who told me that in research, if you're not failing a good percentage of time, you're not doing your job. I don't know what the exact percentage is, but if you don't put in some percentage of your effort at the risk of failure, you will not find the edge of the envelope. You have to define the envelope on what's possible and what isn't. I don't want to see at the end of the year that everything was a success, because then I'll know I didn't really push or define the bounds of the envelope. So I would say that some degree of failure is important to research because I helped define what can be done and what can't be done.

Stern: What gives you the motivation to go forward when you have that failure?

Loce: I don't find that failure is a blow of motivation. I don't give it a second thought. It's like: "Oh, yeah. Well, I found out that's not a fruitful path. Let's try a different solution."

Stern: You just mentioned a manager who gave you advice. Do you have mentors? If so, how do you find a mentor and what role should a mentor play in your life?

Loce: I've had mentors in the past, but it seems in the last few years that I'm the mentor. My early mentors were people who enjoyed invention. They modeled for me how invention is an enjoyable art—whether you are doing the invention or listening to your colleagues' inventions.

One of my mentors was my freshman physics professor. He was also a researcher here at Xerox and later one of my managers. He was a very good writer and communicator. Communication skills help you a lot in moving your ideas forward.

If you invent something and get a patent, you still have to communicate it if you really want to have some sort of impact with it. You've got to communicate it well in writing, and you might have to communicate it well in slideware and in spoken presentation.

My mentor would always try to distill everything down to the message appropriate to a given audience. He taught me how to build the bridge between invention and innovation. If you can invent but you can't sell people on your idea, it's hard to move the invention forward into innovation. If you can champion your ideas and articulate their value and how they work, they will tend to be used much more.

Stern: You said that you are now becoming the mentor. Do you think that happens just as a function of seniority, or is something else at play?

Loce: Well, a few things. Some of it is years. I have been at Xerox thirty-one years, so people sometimes tend to go to the guy with experience. People here know that I have one hundred fifty-seven patents and that I have an official role here as what we call an "intellectual property champion." This role means that people are supposed to come to me when they are having trouble understanding how to go about getting a patent, or want to find out how to better describe their idea so it can be patented. So the attorneys here and the technical community both know that I'm kind of an official mentor in this regard.

But many people, who have been here fifteen years or more, should be taking on somewhat of a mentorship role. In a research lab, even in periods when you are laying off, you'd better be hiring people who have the latest new skills—especially skills that I don't have. A lot of new hires are much better at computer vision and video processing, for example, than I am. But I can help them formulate their ideas into solutions for problems, rather than just staying in their heads or on their desks. A lot of the mentoring I do is also showing them how to communicate their work and to get their solutions out to the community.

Stern: Could you talk about intellectual property?

Loce: There are a lot of approaches to getting intellectual property. There is a patent system, which has a lot of positives to it, but there are some negatives as well. I think the patent system does tend to favor the larger research labs that can put big investments in research. Even though the costs are less for the small inventor, there are some systematic reasons why the corporations that can afford big labs can get more patents.

Bigger corporations often use intellectual property like currency. My company is inventing in crowded areas. If we don't own any intellectual property, but other companies own it all, we have to license everything from them and our products get very expensive. So we want to invent things that are interesting to our business and to our competitors, so we can in effect trade patents across our respective portfolios.

Some critics interpret the statement that we want to patent things that our competitors are interested in as meaning that we want to stop innovation by blocking our competitors from bringing new things to market. No, it's perfectly fine if our competitors bring new things to market. We and our competitors are inventing in a space in which all of us have mutual interest and complementary portfolios of intellectual property. So we use our intellectual property as trading chips to buy our way into the field and maintain our position in it.

Stern: When an individual at your company approaches you with an invention, what sort of advice do you give them?

Loce: I try to get them to understand and state their invention in terms of our company's strategy. I say to them, in effect, "How could we possibly use this? If we couldn't use it, do you know of somebody else outside of Xerox who would use it that we could sell it to?"

So first, is there some value in what you are doing? Once we've figured out what the value is, now let's see if we can articulate exactly what your invention is and how it operates. Is it similar to what anyone else has done? Once you break it down to a crystallized form, it's easier to compare it to other inventions and other known art. When it's in some more nebulous or elaborate form, couched in lots of paragraphs of writing and all sorts of text, it's harder to understand.

But first we try to find its business value either to us or somebody else. Could we sell this? And then we make sure it works. And finally we crystallize it down to its essence and see if we have to worry about prior art.

Stern: Do you get involved at all in the commercialization side? As far as the marketing and the selling?

Loce: I have been at times, but not that often. But I have met with customers, so we can actually talk about the technology. It's enjoyable. Salesmen operate differently. It's a different mode than we operate in. Among scientists, you are often really comfortable talking about all of your mistakes and all of your problems, and this is wrong and that's wrong. But you don't want to go out and start telling customers, "Well, I made this mistake." You've got to edit yourself a little. You don't want tout your product in a biased, over-the-top way, but you don't want to talk the way engineers normally talk. You want to give the real technical information that the person needs to know, not all of your grief in getting there.

Stern: I realize that your work is proprietary, but can you give us a general sense of where your technology and research interests are going?

Loce: Most of it is in video processing and computer vision, with key application fields of transportation, health care, and education. Transportation is probably the easiest one for me to explain without stepping into specifics. Half of the car fuel used in San Francisco is burned looking for a parking space. Thirty to fifty percent of the cars driving around the streets in Brooklyn are looking for parking spaces. By some estimates, you could cut eighteen percent off fuel consumption by reoptimizing traffic lights more intelligently. Some of the transportation problems that we are going after are to save fuel, reduce emissions, reduce congestion, improve highway safety, and reduce the cost of law enforcement.

Solutions have worldwide application. Every country in the world is looking to make their transportation systems more intelligent. Computer vision and video processing are going to be key tools in making highways and transportation systems more intelligent. Look at the cameras that are out there already: there are a million CCTV cameras in London—forty thousand in the London subway system alone. Some of our buses here in Rochester have nine cameras on them.

Stern: Do you have a particular favorite invention?

Loce: It's probably the next one. I'll tell you two of mine that are favorites for different reasons. One was an optical system for a Xerox printer many years ago. It was one of the three main inventions in a printer that contained as many as two hundred patents. That product made about a quarter-billion in annual revenue for a good decade or so. It's satisfying to see something so useful, productive, and lucrative out there.

The other is my very first patent, from 1982. I still kind of like that one. Even though it doesn't look too practical now and I don't think we ever used it, it had a lot of very unusual spatial thinking in it.

Stern: Do you have any general advice that you would offer an inventor?

Loce: Try to surround yourself with like-minded people who really want to build ideas, and don't be afraid of putting half-baked ideas in front of the right team. That really seems to be a very large element of it. Early on when you are inventing, you find that there are some people you can go to and they help you improve your ideas. And there are other people you can go to and they tell you what's wrong. So early on, find those people who you trust will consider your half-baked idea seriously and will try to help you improve it.
They are out there.

Stern: When you are at a dinner party, what do you tell people that you do?

Loce: When I'm at dinner parties, I'm usually talking about mountain climbing.

Stern: What do you do for fun?

Loce: I do like mountain climbing a lot. For my graduation when I finished my PhD in 1993, I climbed a mountain in Mexico: Popocatépetl, which is 17,800 feet. It's a volcano that is now erupting so much, you can't climb it anymore. When my second son was going to be born, my wife said if you're going to go off and do any adventures, do it right now because you're not going to be able to go anywhere for the next couple of years. I went off and climbed a 19,000-foot volcano. Now with my two kids, we're finishing the series of forty-six high peaks in the Adirondacks that we've been working on since my kids were little. So I spend a lot of time in the outdoors: hiking, camping, and various forms of cardio fitness and conditioning. I also spend a lot of time reading literature. Not necessarily technical things. I'm currently working my way through *War and Peace*.

Stern: Do you plan to retire, and what will you do?

Loce: Well, the definition of retirement has to be considered there. I will retire, of course, from my current position at Xerox, but I'm not going to retire from an active life that is conducive to mental health and energy. I don't think I would be mentally healthy if I was not out there engaged in society in some way. It may be less technical, but I definitely plan to be engaged until I drop.

One of the Xerox managers has made the point that if you're working the way you worked five years ago, you're doing something wrong. So whether I'm working in changing ways, or whether I'm home, or whether I'm in some future consulting, university, or volunteer position, I'm hoping that the work will continue changing constantly. My career here has gone from being a technician in the lab, working in the dark with a flashlight, to mentoring a team and guiding people into new technology. And between then and now have come many phases.

Two years ago I was a commencement speaker at the Rochester Institute of Technology, College of Science graduation. I made two points to the graduates. My first point was they had better learn to enjoy lifelong learning, because our society changes so fast that unless you're constantly learning, you're going to be dropped out of the action.

The other point I made was that they should take care to cherish continuity in the midst of change. I told them that my family sold a lot of the land to RIT that would become the university's campus. It was a farm when I was a kid. I learned how to milk a cow and ride a horse on that land. It was almost like an Italian village, where my uncles and aunts lived, and all of our family and our neighbors hung out. All of us who are still alive from those days remain in contact. Well, the faculty at this university are like your aunts and uncles, and the students are like your cousins and siblings, and this place and these people are going to remain an essential part of your community for the rest of your life.

Lonnie Johnson

Engineer
Johnson Research & Development, Inc.

Dr. Lonnie Johnson is a bona fide "rocket scientist" with a mission to save our planet from the consequences of continued dependence on petroleum. Although he has devoted his life to solving some of the world's most complex technological problems, he is best known for his widely popular invention, the Super Soaker water gun. Dr. Johnson is a prolific inventor and holds more than 100 patents, the vast majority of which are energy-related.

In 2011, Dr. Johnson was inducted into the Engineering Hall of Fame by the state of Alabama, becoming the first African American to obtain this honor in the state's history. He was awarded the Breakthrough Award by Popular Mechanics *magazine for an invention that converts heat directly into electricity. He has been featured in numerous magazine and newspaper articles and has appeared on television programs.*

Dr. Johnson holds a bachelor's degree in mechanical engineering, a master's in nuclear engineering, and an honorary doctorate in science from Tuskegee University. He served as a space nuclear power safety officer in the US Air Force, where he helped develop NASA space systems that utilized nuclear power sources. During active-duty assignments in the Air Force and tours with the Jet Propulsion Laboratory (JPL), he helped develop some of the nation's most advanced technological achievements, including the Galileo mission to Jupiter, the Mars Observer Project, the Cassini Mission to Saturn, and the Stealth Bomber (B-2). He has received numerous honors, including two Commendation Medals, the Air Force Achievement Medal, and numerous NASA awards.

Brett Stern: Can you give me your background? Where you were born, your education, and your field of study.

Lonnie Johnson: I was born in Mobile, Alabama. I have a bachelor's in mechanical engineering and a master's in nuclear engineering from Tuskegee University.

Stern: Were you an inventive kid growing up?

Johnson: Oh, absolutely. I wanted to be an engineer before I even knew what an engineer was.

Stern: What were some of the things that you invented or toys that you played with?

Johnson: I assembled a single engine from the parts of two old lawn mower engines that I found at in a scrap heap. Many of the parts, such as the carburetor, were incompatible, but I made them fit anyway. Then I got a scrap go-cart chassis and put the engine on it.

Stern: How old were you?

Johnson: About twelve.

Stern: Were your parents encouraging to you to do these things?

Johnson: Oh, they tolerated me big-time. When I was in high school, I decided to build my own robot as a science project. I started it during my junior year and it took me about a year to build it. I had electronics scattered all over my mother's kitchen. Toward the end of my senior year, there was a science fair sponsored by the University of Alabama. It was the regional conference of schools from all over the Southeast. My robot won first place, so you can imagine how I felt.

Stern: You certainly have a diverse background. Typically, when I interview inventors, they are very focused in their efforts. Could you talk about your various projects and what you're known for?

Johnson: I mentioned that I trained as a nuclear engineer. I've worked at several research laboratories around the country, including the Savannah River National Laboratory and the Oak Ridge National Laboratory. I worked at the Air Force Weapons Laboratory in Sandia, which is in Albuquerque. From there I went to the Jet Propulsion Laboratory in Pasadena, where I worked as a civilian. Then I put the uniform back on and spent some time at Strategic Air Command headquarters in Omaha, Nebraska, and then Edwards Air Force Base in California, where the space shuttle lands.

Stern: What type of work were you doing?

Johnson: I was basically a flight test engineer. My job was to plan the flight test for the stealth bomber and to identify tests needed to make sure that it would meet stealth emission requirements.

At the Savannah River Laboratory, I worked on the plutonium power source that is used on spacecraft. At Oak Ridge National Laboratory, I developed a computer model for the auxiliary cooling system of a nuclear reactor.

Albuquerque is where I really got introduced to the space program. I worked on the Voyager project doing safety analyses, looking at things that could go wrong in the mission to cause a catastrophic event or release of the nuclear power materials into the spacecraft environment. Voyager and Galileo both used nuclear power sources.

I then left the Air Force and went out to JPL, where I worked on the Galileo mission. I was responsible for integrating the power system onto the spacecraft, and I analyzed how much power was needed for each of the functions that the spacecraft had to perform throughout its mission, such as the orbit insertion burns and the science maneuvers.

Stern: Did you get to pick the projects that you worked on?

Johnson: Sometimes. I was hired to work on Galileo. I left the Air Force to work on Galileo because I wanted to know more about interplanetary space craft design. I was assigned to the power systems as a systems engineer, which was a plush position at JPL because the systems team really pulls the whole design together. We got the big picture. I came up with an idea for solving a major problem that my fellow engineers said would not work. Making it work and getting it on the spacecraft was a gratifying moral victory. I felt that I had arrived as an engineer because I was doing things that some of the best engineers in the country said couldn't work. That invention was one of my most significant accomplishments.

Stern: Could you talk about your inventions outside the military?

Johnson: The one that I'm known for is the Super Soaker water gun. Many of the high-performance Nerf guns that Hasbro manufactures are based on my patents as well. When the Super Soaker was successful, I decided that I wanted to go after the other major toy-gun market, which was Nerf. Hasbro was already making them. So, I came up with an entirely new product line, which was much more sophisticated and better-performing than what Hasbro was making at the time.

Stern: You've alternated between working as part of a team and as a lone inventor working at home. Could you talk about the difference?

Johnson: Being part of a team, you are working on larger projects and you have your area of responsibility. With your own projects, you have to solve all problems by yourself. For example, on the Galileo power system project, I was the guy that pushed it. I had to get the right people to buy in and support it, and I had to get the right team on the problem to do the mechanics of making it work, because at JPL, I was not in a hands-on laboratory situation.

I didn't have access to soldering irons, electronic circuits, components, and things like that. I didn't work at that level. I was at the system level where we were actually doing the design and functional requirements for the spacecraft. I had to get other people to do the physical effort of putting the circuit together and demonstrating that it did indeed work.

At home, by contrast, I could do everything myself. In fact, I managed to get some momentum on the power system at JPL by threatening to go home and prototype one of the ideas in my garage. That threat gave me enough leverage to lean on those guys and make the idea work. They didn't want to be outdone by me in my garage.

In general, working at home is very much a matter of deciding what I want to work on and following through on that decision: doing some sketches for the engineering and working out the details from the beginning to the end of the concept. Then I would make all the components in a small machine shop I maintained in my home, assemble them, and test the concept.

Stern: Can you talk about the ideation process that you go through? How you start a project and how you get to the solution?

Johnson: Most of it is initially in my head. One of the things that I consider a gift is that before putting anything on paper, I have the ability to mentally see complex working machines with all the component parts.

Stern: Would you say you're seeing it in 3D or in an exploded view?

Johnson: I can see things in 3D. That's a skill that I have developed. And I've become pretty good at it. Starting as a small child wanting to know how things work and taking toys apart, and the whole nine yards.

Stern: How hands-on do you get once you see this in your head? Do you start sketching? Do you get on the computer and start modeling? Or do you get right into cutting metal and soldering iron?

Johnson: It varies. Every now and then, I get to make devices myself. I make sure that all of the components so that they all fit together dimensionally and work in the end. If I have someone else doing the work, I give them more detail in terms of sketches and dimensions than if I were doing it myself.

Currently, I am fortunate to have a team that does a lot of the mechanics and putting things together, and even solving problems. At this point, I have PhD-level research scientists who help me to accomplish some of the things that I do. Now when I come up with an idea, I can talk to somebody who knows more about it than I do, with whom I can brainstorm and figure out if it's crazy or if it's a great idea.

Stern: Do you have formal brainstorming sessions with people that are working with you?

Johnson: No.

Stern: More in the hallway?

Johnson: Yeah, it's not real structured. We're developing advanced batteries and electrochemistry. Much of the work involves problem solving at the component level. More complex systems as the JTEC engine and things like that would require a lot more sophistication and formality.

Stern: I mean this in a positive way. You're sort of jumping around in different technologies and different fields. What is the incentive to look at such diverse venues? What's going on in your head that wants to look at so many different projects?

Johnson: You mean besides ADD?

Stern: But you are following through on these things, so…

Johnson: Oh, yeah, we follow through and persevere. Some things I've been working on a long time, such as the battery—and even the water gun. I've had many great ideas, but, for a long time, I had trouble getting support and convincing people to back me. I was working on a heat pump when I got the idea for the water gun. I decided that I might as well just go ahead and make the toy, since it was something that anybody could see and appreciate very easily. Then I could use revenue from that to go off and do some of the other things that I want to do. That was the strategy. After the success of the water gun, I returned to working on the heat pump. The JTECH grew out of that effort.

Stern: Would you say there's a difference between being an inventor and being an engineer?

Johnson: Maybe it's a matter of how well the problem is defined. Inventing is more like seeing something beyond what sits in front of you. Once you invent or come up with an idea for something, the engineering is the problem-solving, research, trial and error, and all the other things that you need to do to get to the end. But the initial concept or idea—the seed, if you will—is where the inventive part comes in.

Stern: You said that you're fortunate because your past successes have given you the liberty to develop your next projects. Samuel Morse, who invented the Morse telegraph and the Morse code, was a fine artist and painted portraits to support his telegraphy R&D. It sounds like you're willing to take your past successes and risk their revenue in your next ventures?

Johnson: Yeah. Maybe it's not the wisest thing to do.

Stern: Why do you do that then?

Johnson: That's a good question. When I was at JPL and working on the Galileo spacecraft bound for Jupiter, whenever someone from Washington would come down for some type of review, the justification process they went through was all about the science and all the knowledge we're going to gain, and so forth, and so on. I was an engineer for a spacecraft bound for Jupiter. And you do that because it's a challenge. It's a chance to apply yourself. It's a worthwhile way to spend your life.

Stern: What do you think you are risking? What's the drive?

Johnson: My risk and my drive both come down to confidence, I think. Quite often I will find myself in a situation where things are tough, and I'll wonder if my confidence is justified or, if I have miscalculated. But at the same time, I've had a lot of successes and I know that every ambitious project will have results that are useful, even if it the original goal is not achieved.

Stern: Would you say that you just need to give yourself a tougher problem each time?

Johnson: Seems like it.

Stern: Earlier you said that part of your work at the Air Force Weapons Laboratory was simulating accidents and then analyzing them. It sounds like you were purposely creating failures. What would you say has been your biggest failure so far, and what have you learned from that?

Johnson: Jeez, I don't know. I'm still working on just about every idea I ever had, it seems. So, my biggest failure is not giving up. I fail to give up.

My first patent, which I got back in 1979, was a digital distance-measuring instrument. Basically it was a way of optically reading a binary-coded scale using photo transistors and a magnifying lens. I had the idea, suggested by the operating principle of microfiche, of making a line "one" and the absence of a line "zero," so that I could have a machine read it. That's the base technology that is used in DVDs and CDs. I refer to that idea as "the big fish that got away."

Stern: And why did it get away? What didn't you do?

Johnson: I was in the Air Force at the time and was working on exciting things like Voyager and Galileo, and going to JPL. My day job was providing me a lot of fulfillment—so the inventing was just for fun. I was getting patents just for the heck of it. I wasn't thinking about commercializing the invention at that point.

Stern: It's not that easy to get patents. And they are certainly expensive.

Johnson: Yeah. I actually wrote my first series of patents myself and prose-cuted them through the patent office myself. I learned to do that. I went to the library and I talked to patent attorneys. There was one particular guy who was a patent attorney for the Air Force who tutored me in how to write patents. He offered to write them for me, but I told him no, I wanted to know how to do it myself. I managed to get patents without their costing me so much.

Stern: What do you think makes something unique? What really defines something as new?

Johnson: What defines the level of innovation in a patent application is to some extent subjective on the part of the patent examiners, who often reject a patent because it's not new enough or it's too obvious. In terms of a secret formula for assessing genuine innovation, I don't have one.

Stern: Do you have a particular definition of what an invention is, or what you seek in an invention?

Johnson: It needs to solve a problem and it needs to be useful. Those are my criteria for what I choose to do.

Stern: Where do you seek inspiration for your inventions?

Johnson: Good question. Reading, looking, and figuring out how things work stimulate ideas. I'm very motivated by the prospect of adding any idea to the stock of human endeavor to demonstrate it working, bring it to fruition, and make it available to the world. Part of it also has to do with what I said about the gift that I've been fortunate to receive. I do feel that having skills that some people don't have confers an obligation on me. I can't just squander it. I have to apply it to useful things. More than anything, I guess that motivates me.

Stern: How does that motivation compare to the financial reward?

Johnson: It seems that the financial reward gets squandered. In fact, my fondest dream is to have something to give back and be able to make a difference in the world.

Stern: Do you find you need a certain skill set to be an inventor? What are those skills?

Johnson: In anything you do, clearly the more skills you have, the better off you are going to be. You wouldn't go out and try to be a tennis player if you weren't in good physical condition and you didn't have skills to play tennis.

There are the engineering skills of knowing how things work. Very valuable is the related skill of being able to go into a shop with a set of machines, and make parts, and actually build something and make it work. It's the closest thing to mind over matter that I can think of. You envision what you want an ideal machine to do, and then you fashion all the parts, and you will them to work together in a real machine. It's a very powerful feeling when it works.

Sometimes I'll ask engineers on staff to design something or to get something done, solve a problem, build something, whatever. They go to the computer, do a lot of CAD work, lay out a design, and then they've got to give it to a machinist and somebody who actually knows how to cut metal parts to make it work. At that point, you have a team on the project. And then the real world slips in and they realize that a lot of the stuff that they drew with a sophisticated computer program is not very practical in terms of actual application, so they've got to modify the design, change this, change that, and the end result turns out dramatically different. It's the real world versus the CAD and conceptual world. So, having the skill to bridge those two worlds from the beginning to the end of a problem-solving process is something I consider very valuable.

Stern: Besides the technical side, what about the emotional side?

Johnson: I was going say something about having the business savvy too, because the inventing process involves many contract negotiations and financial responsibilities, but that's the least enjoyable part of my job.

But the emotional side—it's like a thrill ride sometimes. Sometimes it gets fairly scary because you have a lot riding on it and you've invested so heavily in it. In fact, for a long time, I would not accept investments from other people, particularly from friends, because I didn't want to put their resources at risk.

Stern: What made you change that rule?

Johnson: I only change it when I feel things are pretty safe, and even at this point I only have one partner who I have taken real investments from. He is a good friend and I think the world of him. I would not have done that if I didn't feel like success was pretty much in hand and that I'd taken the really high risk off.

Stern: Are there any inventions or inventors that you admire?

Johnson: Thomas Edison and George Washington Carver were my heroes growing up, but my biggest hero was my dad. He was not an inventor or scientist, but he used to repair things around the house and I would always get excited when he worked on the car or did something like that. I always had to see how it was done and what he was doing, so he was my role model.

Stern: Do you have a favorite product or technology that you use all the time that you really admire?

Johnson: You mean besides my laptop? I guess it would have to be my smartphone because I listen to music on it.

Stern: Do you have any advice for would-be inventors out there?

Johnson: I guess the main advice I would give people is to have more than one idea. It puts the odds in your favor, because you don't know which one is going to succeed. There are many variables that you really cannot control.

Stern: When you're at a dinner party, what do you tell people that you do?

Johnson: Usually I don't.

Stern: What do you do for fun?

Johnson: I like to swim, but more than anything just hanging out with my wife and the kids. Fortunately or unfortunately, the thing that I did for fun as a hobby became my job, so my hobby went away. I enjoy music, I enjoy parties, and I enjoy the family.

Stern: Do you plan to retire?

Johnson: I don't think so. When I'm finished with some of the things I'm working on now, I think I will take on some easier projects that won't take so long to solve and maybe won't be as stressful.

Stern: Any final advice you want to give to would-be inventors out there?

Johnson: Just persevere.

Tim Leatherman

Co-Founder
Leatherman Tool Group, Inc.

Tim Leatherman was born in Portland, Oregon, and is a mechanical engineering graduate of Oregon State University. Leatherman was inspired to design his folding tool while he and his wife traveled overseas in 1975, often attempting to use a simple pocketknife to repair their repeatedly malfunctioning car or leaky hotel plumbing. He spent several years perfecting the prototype and received his first US patent (4,238,862) in 1980.

Leatherman Tool Group, Inc. Was formed in 1983. Its first product, the Pocket Survival Tool initially sold through Cabela's and Early Winters' mail-order catalogs. Leatherman sold nearly 30,000 tools in 1984, fueling additional product development and both rapid company growth and manufacturing capacity. Today, Leatherman's only manufacturing facility is located in Portland, Oregon.

In 2007, Leatherman was inducted into the Blade magazine Cutlery Hall of Fame in recognition of his design impact on cutlery history.

Leatherman Tool Group also manufactures a line of multi-tools designed specifically for military and law-enforcement personnel, as well as accessories for carrying and expanding the function of its tools. Currently, the company produces 46 products, which are sold in over 110 countries.

Brett Stern: Can you tell me your background? Where you grew up and what kind of education you had?

Tim Leatherman: I was born and raised in Portland, Oregon, and went to university at Oregon State. I got a degree in mechanical engineering. After graduation, I worked for a year for a company that designed oil refineries.

My job was to sit at a drafting table in a room that had about two hundred other drafting tables, and draw lines on a piece of paper where pipes were supposed to go in an oil refinery. That was in Los Angeles. In that whole year, I never once saw a real oil refinery. So I couldn't stand that anymore. I took a leave, came back to Portland, kicked around and did some construction work while my girlfriend finished university. When she finished, I followed her home to her country. We got married there.

Stern: What country was that?

Leatherman: That was Vietnam. We were there for two-and-a-half years. Right before the Communists took over in 1975, we left, came back to Portland, and got her family settled. My wife and I then went to Europe on the famous trip where I thought of the idea for the Leatherman Tool. We came back nine months later, and I've been here ever since.

Stern: Growing up as a kid, did you consider yourself mechanically inclined?

Leatherman: I would say no. I would say my brother was far more mechanically inclined than I was. He was always tinkering with machines and trying to build motorized go-karts and stuff like that. But I was not so mechanically inclined until I got to Vietnam.

One of my jobs there was teaching English and one of the other English teachers was a sort of a professorial type, almost to the point of being a stereotype. He smoked a pipe and was klutzy. He and I were talking one day and he said, "You know, motor scooters are really pretty simple." I thought, if Clyde Correll thinks that a motor scooter is simple, then there's something wrong with me. It was basically that statement that spurred me into taking a keen interest in the way things work.

Stern: Yet in school, you studied mechanical engineering.

Leatherman: I went to school to pass the exams. The curriculum was tough. I had to take a lot of hours every term to graduate in four years. And I was paying my own way. I pretty much just took the classes, listened to the lectures, studied the material, did the homework, and followed the procedure that the textbook said to solve the problems.

I turned in the work and hoped the exams were nearly the same kinds of things they asked us to do for homework. That's how I managed to pass the exams. If one of the test questions was to calculate the diameter of a cable that it takes to lift a salvaged automobile at a wrecking yard, I never did stop to think an automobile weighed four thousand pounds and to actually visualize it. If the answer turned out to be "three-sixteenth-inch cable," I never stopped to really let that knowledge implant in my mind.

Stern: Could you describe the product that you've invented? What you're known for?

Leatherman: The product I'm famous for is what's now called the Leatherman Tool. We originally called it the Pocket Survival Tool. As far as the background for that—how the idea came to me—it was when my wife and I took a budget trip to Europe in 1975. I bought a used car in Amsterdam for $300 and we stayed in cheap hotels and pensions. I carried a pocket knife and I often needed pliers.

Stern: To fix the car?

Leatherman: To fix the car, and even things like fixing the hotel plumbing. And that trip was sort of a "what are we going to do with the rest of our lives?" trip. I jotted down notes on a piece of paper that I kept in my shirt pocket, and one of the things I jotted down, in essence, was "put a pair of pliers in a pocketknife."

Stern: What were some of the other things that were going to be your life directions?

Leatherman: By then, I was pretty keen on the way things work and fixing things. One of my thoughts was to cluster repair shops into a kind of shopping center. A place where all types of different repairs were done.

Sort of the concept of an Asian market, where if you go to buy a cloth, all the cloth vendors are in one section. I've looked at that list over the years, and there were several things that have come to pass.

Stern: So you still have the list, the actual piece of paper?

Leatherman: Yeah, I still have that piece of paper. There was an idea on there to automate postage and to read addresses, something like a bar scanner.

Stern: After you come back from the trip, you look at the list. Why did you pick the pocketknife with pliers?

Leatherman: I don't know. For some reason, that was the idea that appealed to me the most to work on. I asked my wife if I could build one, just one, just for me. She asked how long it would take. I said maybe a month. She said okay. So she went to work to support us, and I went to the garage and picked up a file and a hacksaw, and I tried to build what was in my mind. Three years later, I had a prototype I liked.

Stern: How many prototypes did it take to get to the one that you liked?

Leatherman: Many. And several fundamentally different, evolutionary different designs. And then hundreds of broken parts and failed attempts.

Stern: Obviously, you have a background in mechanical engineering. You knew how to draft. Were you using those skills?

Leatherman: Basically, I was building it by cut-and-try. I built first, and then if I came up with anything I liked, I drew it afterward. I wasn't really engineering it. I wasn't calculating forces or anything like that.

Stern: You were more of a tinkerer.

Leatherman: Yes. I just built it. Just built it to see if it worked.

Stern: During those three years, this was, in a sense, your full-time job.

Leatherman: Yes. A full-time job. With no income.

Stern: What was in your head during that process: one day you're going to get a prototype that works? What was the next step that you thought you'd do?

Leatherman: Well, from the very beginning I thought that if I liked it, that there might be a market for it. So I thought that maybe there would be some way I could make some money off the idea if I built one that I liked.

Stern: So the thought was that you would start a business?

Leatherman: The first thought was probably like every inventor's. I placed a value of exactly one million dollars on my invention. It's funny how that hasn't changed with inflation Today's inventors also think their idea is worth one million dollars. I originally thought that I could design the product, build the prototype, patent it, take the patent and prototype to a knife manufacturer, they would pay me a million dollars, and I could sit back and live happily ever after.

Stern: How did that go?

Leatherman: Not well.

Stern: You did patent the idea.

Leatherman: Yes. I got to the point where I had a prototype I liked and I filed for a patent. At that point, I felt confident enough to try to license or sell it. I think it was around eighteen months between the time I filed till the time I got the patent. Prior to filing the patent, I was teaching myself as I went along. I went to the library and learned as much as I could about patents. I even started trying to write my own patent, and then finally got to the point—and succeeded to the point—to where I got to the claims. I decided at that point, it was really worth it to have a professional write the claims.

Stern: What was the state of the art in pocket knives with tools in them at the time?

Leatherman: As it turned out, after we filed the patent, the patent office said that the state of the art included scissors that had folding handles. And therefore, it said that the basic concept of having a pair of pliers with folding handles had already been covered, because scissors, too, have a pivot point in the middle.

We tried to argue back that the pliers made it fundamentally different—that a design strong enough to transmit force to pliers as opposed to scissors required new and innovative features. But the patent office declined our amendments, our petitions. In that original prototype, I also had a clamping device and two kinds of pliers in the tool. I had needle-nose pliers and regular pliers. The needle-nose pliers could be independent of, but driven by, the regular pliers. We got full coverage on those claims. I also had a clamping feature where I could engage something between either pair of jaws, and the object would stay clamped when you let go of the handles.

Stern: So you have a patent and prototypes in hand. What was the process then?

Leatherman: Believing that I had a knife with a pair of pliers, the process was to go to a knife company and show them my new pocket knife. I started local. There was a company in town called Gerber, so I went to them. They were quite gracious. They borrowed my prototype. They showed it to a retailer. They studied it themselves. A week or two later, I checked back with them, and they said, "Thanks. Sorry. We appreciate you showing this to us, but it's not a knife, it's a tool."

Stern: And they are knife makers, not toolmakers.

Leatherman: Yes. I still thought I had a knife with pliers. I went to other local knife companies, and then eventually went national—and even international by correspondence. And all of them said no. This was about the third or fourth year. Not being a total idiot, I got the message and said, "Okay, if they're calling it a tool, then I'll go to the tool companies." But the tool companies said, "It's not a tool, it's a gadget. Gadgets don't sell."

At that point, my wife and I had a talk. We decided that I should get a real job while still trying to pursue my dream of getting this tool to the market.

I'm sure I had an opportunity to go back into engineering. I was almost 99.9 percent positive I could have gotten a job with one of the companies that I showed the invention to, working for them as an engineer. But I still had the dream of getting this tool on the market. But it was looking more and more like to do it, I'd have to start my own business.

Stern: What was the motivation? To prove your innovation skill? Financial? Intellectual? What do you think was the drive behind it?

Leatherman: I think it was just stubbornness and perseverance. I think when I started out, I established a goal and I was just too stubborn to give up without achieving that goal.

Stern: Certainly a common trait amongst inventors. Would you say that this was part of your persona prior to this project, or were you growing into being this stubborn person?

Leatherman: Probably growing into it. There might be a little bit of evidence prior to this project that if I decided to do something, I'd stick to it until I achieved it.

Stern: You woke up and said, "Okay, I'm going to start a business doing this because no one would license the patent." What was the process then?

Leatherman: No, it wasn't quite that simple. I looked in the want ads and found someone advertising a position for a salesman, an outside salesman selling welding products. I applied for that and decided to take it. I knew that if I was going to get into business, the ability to sell was important. I also knew that I knew nothing about business, nothing about invoices or…

Stern: So this was your education then?

Leatherman: Yeah. This was a way of training myself to still achieve my goal. The job also had the benefit of quite a bit of freedom because the headquarters of this company were back in New York—and I had a territory in southwest Washington. As long as I met my quota, they left me alone. I had quite a bit of freedom to manage my own time. So my day job was selling welding products, and my second job was trying to get my pocket knife on the market.

The welding products job also gave me some travel opportunities. I was able to visit a few in companies in person—some knife companies back East—and continue trying to sell the patent rights. I still failed. Finally, after about four years of the second job and seven years from the beginning, I had just about given up. But a friend of mine from my college days named Steve Berliner had been following everything all this time. He stepped in and said, "There are still a few things you haven't tried." We established a 75/25 partnership and pursued some more things.

At that point, we had pretty much decided that if we're going to get the tool on the market, we'd have to start our own business. But we thought if we're going to start a business, then we should have an order for some tools before we commit to the cost of getting into production. Among our first thoughts was to sell to the Army. If the Army bought and issued one to every soldier, that would be a good order. But the Army said no.

Then we thought about AT&T. This was before the telephone company broke up and they still had tens of thousands of people repairing telephones. We thought it would be a lot easier to carry our compact tool instead of that big tool belt the repairmen have around their waists. But AT&T said no.

Then we thought mail-order catalogs. We contacted most of them by letter, but one of them by telephone. It was a company called Early Winters in Seattle. They gave us an appointment to come up and see them, and so we drove up. Steve and I had done a little bit of evaluation of what it would cost to get into business. We thought that if we could make and sell four thousand tools in our first year, we could pay back our investment. But even at that, we knew we would pay ourselves nothing for our time. So we were looking for an order for two thousand of the four thousand tools.

At Early Winters we showed them my prototype. They said, "It looks interesting. How much?" I hadn't exactly decided the price.

Stern: You were so used to people saying no.

Leatherman: Yes. I thought quickly and said, "Forty dollars." And they said, "Hmm, that means we have to sell it for $80. Sorry, we don't think people will pay that much." But they did us a big, big favor. Instead of saying go away, don't let the door hit you on your way out, they said, "Sit down. Let's take a look at this."

They asked how could I make it less expensive. I said, "Well, the first thing I would do is take out the clamping feature, 'cause if anything were to break, that would be the first thing." They said to take it out, and then they asked, "What else?"

I said, "Well, the two kinds of pliers—the needle-nose independent of and driven by the regular pliers is very complex and very expensive to make." They said, "Take it out. Just have one pair of pliers, have the needles at the end, regular in the middle, and your wire cutters down at the base. Okay, what else can you do to make the tool less expensive?"

I said, "I really admire the Swiss. I had no idea how difficult it was to put scissors in a small tool or knife until I tried to do it myself." They said "Take it out. Substitute it with a screwdriver or something."

Stern: Did you feel that you were taking out all the good things?

Leatherman: In a way. But I came back to Portland and went back to the garage. About six months later, I had a new prototype. We called Early Winters. They gave us an appointment. We came up and showed it to them.
This prototype looked very close to the way the original Leatherman tool—the Pocket Survival Tool—ended up looking.

Stern: What was as the price point?

Leatherman: Twenty-four dollars. They said they liked the tool. We said, "Well then, you're ready to order two thousand?" And they said, "No." So with that, we came back to Portland, took a picture of the tool, created a brochure, wrote a bunch of cover letters, and re-solicited the mail-order catalogs with the new prototype. A catalog company named Cabela's responded with a purchase order for five hundred tools. Finally, after eight years, I had an order.

Steve and I both realized that this was only five hundred tools. It wasn't two thousand. But on the basis of that order, we decided to go into production. Luckily, Steve's dad had a metal-working business, so we were able to take the equipment I had accumulated in my garage, move it into his dad's business, and then his dad let us use his employees and his equipment on a subcontract basis.

The Cabela's order was in late May of 1983, but they wanted delivery in late December of 1983. We thought that seven months was plenty of time to get into production. Still thinking we had a knife with a pair of pliers, and thinking that it would appeal to knife enthusiasts, we found that there were some small publications that were directed toward knife enthusiasts and knife collectors. There were about thirty thousand of these people, and we placed an ad to sell the tool direct for $39.95. While we were in production, we got a call from Early Winters. They said, "Put us down for two hundred fifty, also for delivery in late December." So the seven months went very quickly.

Stern: You're the president of the business, but are you in the shop?

Leatherman: Yeah, I'm the president. Steve and I had sort of a division of labor where he was more on the business side, and I was more on the production side.

About three days after Christmas of 1983, we got a call from Early Winters. They were asking about their two hundred fifty tools: where are they? I said, "We're a little late, but we'll have them for you any day now." They said, "We need them really badly because they've all been sold. And here's an order for five hundred more." A week after that, they called and said the five hundred were gone. "Here's an order for seven hundred fifty." Two weeks later, the seven hundred fifty were sold. "Here's an order for one thousand."

Meanwhile, Cabela's was selling their five hundred and placed reorders. It turned out that all the catalogs were monitoring each other, and if they got a sense that any product was a hot product, then they'd all jumped on the bandwagon. All the catalogs that had turned us down prior to this started calling. Of course, we were smart enough to not say anything like, "Why did you turn us down?" Now we were getting orders from L.L. Bean and Eddie Bauer.

Stern: You're still in the shop at this point?

Leatherman: Yeah, we're still in Steve's dad's shop. We had elbowed our way into about one thousand square feet of space. We had about four people doing production work.

Stern: Was this your full-time job now?

Leatherman: Actually, no. I kept my welding sales job through March or April of 1984. Today, we have brought a lot more work in-house, but back then we were making the two handles from scratch, making two of the blades from scratch, and having some blades subcontracted. We did manage to ship about two hundred tools in the last few days of 1983.

In that first full year, we were hoping to make and sell four thousand tools. Business was so good that we were actually able to make and sell almost thirty thousand tools. The year after that, seventy thousand tools. Then it just grew exponentially until in 1993, ten years later, we were making and selling one million of these tools a year.

Stern: At what point did you realize that this was a business?

Leatherman: I think it was on that third call from Early Winters, when they said the seven hundred fifty are gone, here's an order for one thousand.

Stern: Whom do you think originally were the people using the product?

Leatherman: They were primarily outdoors people. They were what we now call "outdoor recreationists," the whole gamut from hunters to fishermen to bird watchers. Then there was another group that we called "professional users"—tradesmen, including the military, people who used the tool on the job. And then the do-it-yourselfers. People would buy the tool and keep it in the glove box of their automobile.

Stern: Mine's in my glove box.

Leatherman: Or keep them in their kitchen drawer at home. There was also a fourth category of customers—gift buyers who would buy for those three other categories. It turned out that the market saw this not as a knife with a pair of pliers, but as a pocket tool with a knife blade, which put it into a whole new product category.

At that point, we knew we had a tiger by the tail.

Stern: You had just one product. At what point did you start expanding on that product line?

Leatherman: Fairly early. We were told that it was impossible to be a successful company with just one product. Within a couple of years, we decided we should add another product to the line. Steve and I talked it over, trying to decide whether to go bigger or go smaller.

We decided to go smaller. I went out in the shop and pretty much repeated the same process as before. We had a concept. I had some design ideas. I started building parts and putting stuff together. About six months later, I came back with what became the Leatherman Mini-Tool. We put it in production and offered it for sale. It sold steadily, but never spectacularly. It was never more than five percent of our sales.

Meanwhile, the original Pocket Survival Tool was zooming up to one million units per year. So it wasn't until 1994 that we considered adding a third new product. At that point we had employees, lots of employees, even engineers working for us—including a bright, young engineer that we had just hired right out of Oregon State named Ben Rivera. He was working for us as a manufacturing engineer, but we quickly found out that he had an aptitude for product design. The result of going bigger was the Super Tool, which within three years became a million-unit-seller in a year.

Stern: When did it go from a job shop to becoming a business?

Leatherman: I disagree that we were a job shop in the beginning because that would indicate that we were doing work for others. Right from the start, we were just making Leatherman tools. Our business was called Leatherman Tool Group, Inc. We picked the name Leatherman Tool Group because we wanted to make ourselves sound bigger than we were. But in the beginning, it was just Steve and me, and a few of his dad's employees working for us on a subcontract basis. However, we were creating, making, selling, and collecting—all the four essentials of being in business.

Stern: How do you think the tool has influenced society?

Leatherman: Based on the testimonials we have had, it's done everything from saving people steps, to saving people trips, to saving lives. I feel very proud of that. But the thing I'm most proud about is the jobs created. The tool fell into a whole new product category that had never existed before. These were new jobs. We weren't taking away from anybody else's work. It has grown to over three hundred jobs. I was, and still am, most proud of having created jobs.

Stern: During this process, there were certainly mistakes that you made. Any really good failures that you can think of? What's the biggest failure?

Leatherman: The first big mistake was to use a potential competitor as a subcontractor. From the very beginning, we did not do our own heat treating. We always jobbed that out. As we grew and built our own building, we always intended to bring it in-house, but we kept growing so fast that the space reserved for heat treating always got taken up with production. So to this day, we still have our heat treating done outside. I used Gerber to heat treat our blades, so of course they knew our sales volumes. I think they were surprised that the volumes were so high. They became our first competitor.

Stern: You gave them an inside spy into your system.

Leatherman: Yeah. I gave them an insight into how fast things were growing. There was one year when we planned to double our production and we had commitments for that doubled production by March 31. We were turning customers down. I remember Kmart came to us at a trade show and said, "We'd like to be ordering from you." We said we'd be happy to take your order and we'll ship when we say we will ship, but you will have to place your orders six months in advance. They said, "No, we don't work that way." We said great, that's fine. Eventually we'll increase enough that we will be able to ship within a much more reasonable delivery time, and when that time comes, we'll be happy to work with you.

Meanwhile, Gerber was already selling their knives to Kmart and when they went in with their tool, of course Kmart said, "Sure, we'll take it." There were a couple of other mistakes. We grew very fast, but we should have grown even faster. We allowed competitors to get a toehold.

Another mistake was being unaware that California had a much more stringent "Made in USA" law than the federal government did. Finding out that we unknowingly violated it, and having a class action lawsuit filed against us resulted in the class getting nothing, but the lawyers getting millions.

Stern: You're obviously committed to making product domestically. Have you had the opportunity to outsource the production?

Leatherman: Sure There's a town in China where almost all the knockoffs are made. I've been there and even been welcomed in their factories. But I'm committed to doing production domestically, as much of it as reasonably possible, because it goes back to being proud of the jobs created and wanting to keep as many of those jobs here as we can.

Stern: Do you think manufacturing is coming back?

Leatherman: Not necessarily. It may. I think there are always companies that are going to be looking for the low-cost option, and I think there are still lots of places where the low-cost option can move. When I was younger, Japan had a reputation for making stuff cheap. Then it moved to Korea, and then Taiwan, and then China. Now it's moving to other places around the world.

Stern: Do you think the consumer appreciates something made in America?

Leatherman: I think they say they do, but I think when it actually comes to where they put their dollars, some, including me, just can't resist the allure of a cheap product; at least initially. I don't know if you remember this, but maybe twenty or thirty years ago, America was being flooded with socket sets. Thirty piece socket sets—$2.95 for thirty pieces. You just can't resist it. Even I bought them. And I think what happens is you buy it and you realize that the concept is good, but the quality is terrible. So then the next time, instead of buying nine more and spending $30 for ten socket sets, and still needing another, you buy Snap-On, or Craftsman, or another good brand, and you just do it once and they last your lifetime.

I think the same thing has happened with Leatherman tools. Potential customers have gone into a store and they see a Leatherman tool under lock and key in the glass cabinet for $70. Then they see a Chinese knockoff in a blister pack on a peg for $6.95—you just can't help but buy that first. Besides, they can't even find the clerk to unlock the cabinet. So they buy the Chinese tool, realize the concept is good but the quality is terrible, and then they go back and buy ours.

Stern: Let's go back to your failures and back to when you taught yourself how to be a businessperson. Have you had mentors during your career? And where did you find those people?

Leatherman: Well, my very first mentor was in the eight-year period when I was building my prototypes. It turned out my brother-in-law was a machinist, a skilled machinist. He had some equipment and I was able to do my work in his garage. He taught me enough—by no means enough to be a skilled machinist—but enough to do what I needed to do to make my prototypes.

Stern: During the developmental process, where do you seek inspiration?

Leatherman: Well, the initial idea comes from a need. But thinking back to the eight years in the garage, during the development process I got inspiration from anything and everything that I could put my hands on, read about, or I could see. Anything in the way of tools or knives.

Stern: In the marketplace?

Leatherman: Yeah, for design ideas. And I guess that's still pretty much true. Although since we've been in business, we've gotten feedback from our customers, who, for example, tell us that our tool is wonderful, but if it were bigger and stronger and more heavy-duty, it would be even better. Or, "your tool is wonderful, but I also wish it had a saw blade." And, "your tool is wonderful, but I wished you hadn't taken those scissors out. I want you to put those back in." And, "your tool is wonderful, but I occasionally open a bottle of wine and I'd like to have a corkscrew." So we have all this feedback from customers, which has greatly driven product development.

Stern: Do you consider yourself an inventor?

Leatherman: Yes.

Stern: What skill sets do you think a person needs to be an inventor?

Leatherman: I think that you need a certain amount of creativity. You need a lot of persistence. I think that you learn far more from building your own prototypes than you do from jobbing it out. I personally believe that it's key that if you come up with an idea and you want to develop it into an invention, that if you can't build a prototype, then don't do it.

Stern: How do you define creativity?

Leatherman: I'd say creativity is coming up with a way of doing something that hasn't been done before.

Stern: Problem solving?

Leatherman: Yes. Solving a problem.

Stern: Unrelated to the products that you are doing here, do you have a favorite product or a favorite item that you like just having around? That you keep on your desk or in your glove compartment?

Leatherman: Well, I think the greatest invention of all time—the invention that did the most to help the most people and improve the quality of life for the most people—are eyeglasses.

Stern: At what age did you start wearing eyeglasses?

Leatherman: About my mid-forties, when my arms got too short to be able to read the newspaper.

Stern: Are there any inventions that you've thought of related to the eyeglasses or anything?

Leatherman: No, I haven't.

But I think the three best inventions fabricated out of metal—that are tool-type inventions—are the Crescent wrench, or an adjustable wrench to use its generic name, the Vise-Grip, or locking pliers to use its generic name, and the little Trim brand fingernail clippers. And I think the Leatherman Tool might have a place in the top four or five tools.

Stern: When you were running the business day-to-day, how did you motivate people?

Leatherman: Several things. I think the personal touch is very important. I used to go through the shop and greet every employee by name every day, up till we had about two hundred fifty employees—when I realized that a minute talking with each employee was four hours a day.

Stern: Do you think they appreciated it?

Leatherman: Yes. I remember telling an employee, "When I hired you I didn't just get your hands, I also got your mind—and I'll gladly hear any idea you have about how you can do your job better and how we can make the company more successful."

Stern: Do you get a lot of feedback from employees in production about how to do things better?

Leatherman: Yeah, a lot of things. During that time when we were ramping up production—when we were going from two hundred to thirty thousand to seventy thousand tools per year—a lot of the ways of doing that came from the employees themselves.

Stern: Do you have a next project that you are working on here at the company?

Leatherman: Not one particularly, but the company always has things. We are working on product ideas three years out. And we are always open to receiving ideas from outside.

Stern: Do you take ideas if they are patented?

Leatherman: Yes. In fact we prefer that they are patented. That way there won't be any misunderstanding as to who thought of the idea and the invention that resulted from the idea.

Stern: When I say I'm interviewing Tim Leatherman of Leatherman Tools, people are inquisitive. They just don't realize that there is actually a guy named Leatherman. They thought it was just sort of a generic name or term. Would you say that's true?

Leatherman: Definitely true. We get many, many people saying that they thought we were a one hundred-year-old company that started out making leather products. They had no idea that there was actually someone named Leatherman that started the company—and only twenty-eight years ago.

Stern: Have you ever thought of doing leather tools?

Leatherman: We have a leather punch in our tool.

Stern: When you're at a dinner party, what do you tell people that you do?

Leatherman: If I'm in a playful mood, I tell people I'm an inventor and usually that attracts a follow-up question. Then I say I started a company that makes a new kind of pocket knife, which I'll describe and which they'll recognize as the Leatherman Tool.

In the work environment you can often spot the owner doing the simplest things. I remember once going to a movie theater and thinking maybe that the guy standing behind the counter serving the popcorn was also the owner, so I asked, and he kind of sheepishly said, "Well, yes, but I do everything. I'm the chief cook and bottle washer." Sometimes that's how I describe my job, too. You're leading, but you're also serving your employees and your customers.

Stern: Do you consider yourself retired now?

Leatherman: I consider myself eighty percent retired. But I'm still the majority owner and still chairman. And, I still do some work at the company. Public relations–type work. And I'm quite happy to represent the company when traveling internationally, to international distributors. And I do what I can while I'm there to help sell tools.

Stern: Your product is selling all over the world?

Leatherman: Yes. We have sales in more than eighty countries around the world. About thirty percent of our sales are export sales.

Stern: What countries are the top three for exports?

Leatherman: It varies from year to year, but usually the countries vying for the top three are England, Germany, Australia, and amazingly, South Africa.

Stern: Are there products specifically for those export countries, or are they all out of the same box?

Leatherman: No, they're pretty much the same. We definitely think about whether we should be modifying our tools to particular markets, but so far our tools sell pretty well everywhere without modification.

Stern: What do you do for fun?

Leatherman: I play a lot of tennis. Go to movies. Read. Travel. Get to sleep more. That's nice.

Stern: Any countries that you haven't been to that you want to go to?

Leatherman: Yeah, I haven't been to the Middle East. I'd like to go there. There are several countries in Asia I'd still like to see. I haven't been to Egypt. I haven't been to Portugal. I haven't been to Peru. I'd like to see Machu Picchu.

Stern: In looking back at this list that you did when you were traveling thirty-plus years ago, do you feel that you've been successful in what you did?

Leatherman: Definitely. I feel successful.

Stern: How has success changed your life?

Leatherman: I don't think it's changed it a real lot. I live comfortably, but still fairly frugally. I hope I'm still a decent human being. I pretty much do the same things I've always done.

Stern: Any final words you want to say on how people should live their lives, or invent things, or be innovative?

Leatherman: To prospective inventors I say rule one is perseverance. If you are working on an invention, at some point you are probably going to think about giving up and going on to something else. I realize there is a fine line between perseverance and failure to accept reality, but I recommend you lean toward staying with it too long, rather than giving up too soon.

Advice number two: be willing to learn what you need to know as you go along. This includes learning the skills you need to build your own prototype, and write and draw your own patent.

Number three: if you want to be a successful inventor—to get your product on the market and make some money from it—you need to be willing to change the product even if you love it, but no one will buy it. You will have to do market research and cost studies. You will have to be willing to go into business for yourself to produce and sell your product if no one else will buy your patent from you. Don't fall for the front-money phonies—the people who advertise they will help you sell your invention if you give them money up front. The best sales person for your invention is you, the inventor.

Reyn Guyer

Toy Inventor

Reyn Guyer Creative Group

Reyn Guyer *is the developer and inventor of the game,* Twister, *which was licensed to the Milton Bradley Company, as well as the Nerf Ball and its spinoffs, which were licensed to Parker Brothers.*

In 1986, with his daughter Ree, Guyer founded Nashville's Wrensong Music, which has won two CMA (Country Music Association) Song of the Year awards and charted several hundred top-ten songs. He founded and chairs Winsor Learning, which remediates children's reading skills worldwide. He cowrote the musical Stained Glass, *which is currently setting up in theaters around the country. Guyer's most recent project is a lawn game called* KingsCourt, *which he invented and is marketing.*

Brett Stern: What's your background? Where you were born, where were you educated, and what was your field of study?

Reyn Guyer: I was born in St. Paul, Minnesota. I attended a country day school by the name of St. Paul Academy. When I studied for my BA at Dartmouth, I was an English major with the intent to be a journalist.

Stern: Were you inventing things growing up?

Guyer: To some extent. I'm not an engineer type, and my math skills are very inadequate for algebra and that kind of thing. I'm very good intuitively at numbers, and so it's an odd combination.

Stern: How did you get involved in toy invention?

Guyer: Well, instead of becoming a journalist, my father enlisted me to try his business, which he had just started six months prior to my graduating from college. It was a design business, and the objectives were twofold. One to help *Fortune* 500 companies in the Midwest with their in-store display materials. The other objective was to create new package design in paper carton packaging.

My father, over his career with Waldorf Paper Products Company, which was a prominent paper box company in the Midwest, had his name on at least one hundred twenty patents in the paper carton field. He had aspirations to go on and do this independent of the Waldorf Paper Company, and hope that he could invent some new things. He and a fellow by the name of Carl Anderson did extensive work to develop new cigarette carton packages, three of which I wish I had in some way saved. They were so unique and so practical. I wish I had them now.

But anyway, the company had those two intents to develop display materials for *Fortune* 500 companies and also to do design work for companies. By example, we did all of the coordinating of the packaging for Land O'Lakes products across the country. And he designed new logos for people like US Gypsum. We did display materials—I'm just picking some names—for Pillsbury, Kraft Foods, Brach's candies, Johnsons Wax, 3 M, Scotch Tape. So we had a wide variety of *Fortune* 500 companies whom we helped with their display materials.

Stern: How then did you go from the design side to the toy-inventing side?

Guyer: The one thing that was built into the business was that, to be paid for our work, we had to first develop prototypes of the product that we were going to display. We did not have a product that was a widget of some sort, or something we were stamping out or manufacturing. Each challenge thrown to us by the companies required that we speculate on developing a display that would be better than our competition's. We were accustomed to working on that basis, which is to develop the product first and then to sell it.

It's a different ethic than having a product and the challenge is to market it, and advertise it, and vend it. We were always trying to invent something new or better and take it to our client. If they liked it, we would then arrange to be paid for it. That was the ethic and the manner of conducting business that I was used to. That being the case, I also had some doubt as to the long-term profitability of a company that was doing what we were doing—developing display materials and new packages.

Stern: Was it based too much on speculative work, and you didn't know if you were going to get paid for the ideas?

Guyer: Well, if they didn't choose our design or our idea, we were not paid for it. So everything was a speculation. It became apparent to me, having been in the business for eight or nine years, and having pretty much taken over the reins of the business because my father was beginning to back away, that I should be looking for some form of alternate operations. Some other kind of business concept that would perhaps be more profitable and a longer-term opportunity than the one that we were engaged in.

I was always looking for something. I tried a lot of things. One of the ideas, by example, was Pizza in the Round. At that point, there were just a couple of what were then called "radar ovens," now microwave ovens, in town. One was at the Star Tribune building in Minneapolis. Pizza was working its way West from Boston, where I had become familiar with it. There were only a couple of places in Twin Cities that made a good pizza. So I had it in mind that we could freeze the dough and apply the sauces, the cheeses, and the tomato sauce, and the additions to that in a very short space of time using the radar ovens.

I engaged one of the leading architects in the Twin Cities to come up with design for a restaurant, which was very clever. It looked very Venetian with poles that had barber-pole wraparounds. It was a good idea, but I backed away from it. I was fairly far down the line when I realized I didn't know anything about the restaurant business.

Stern: At this point, you are just sitting around brainstorming ideas, looking for new businesses to go into?

Guyer: Well, I'm not sure I was sitting around, but I was always looking for them, for new alternatives. Yes.

Stern: Eventually you got involved in the toy business.

Guyer: There again, we backed into it. I was working on an idea for a shoe polish promotion for one of our clients. We would make a display that would sell the shoe polish. But they also wanted what was known as a self-liquidating premium. Something where you added a dollar to the label and sent it in, and you got something for that.

I was trying to develop the "something," and I came up with this idea. I thought, since it was a kids' back-to-school promotion, shining the kids' shoes—which seems rather odd in this day and age, but that's what was done by mothers, attentive mothers, at that time—and I thought, "Well, let's see. What if we put something colored and wrapped a piece of paper around the kids' ankles that would be different colors." Then we have this grid that's on the floor, just a large paper sheet, we could have the kids move around and try to get across to the other side in some way, taking turns moving and trying to block the other person's move. And all of a sudden, I was sitting at my desk, and—"Wait a minute! This is a lot bigger than that. *The people are the players*!"

I went out into the studio and got some of the artists and some of the assistants and secretaries, and I gathered eight people on this grid. I drew twenty-four one-foot squares on the grid that was six feet by four feet. And I made two of the people the blue team, and two of the people were the yellow team, and two were the red team, and two were the green team. I positioned them in the corners and said, "Now your job is to get to the other side when it's your turn to move." Well, of course, on the short notice, it wasn't what you'd call a well-thought-out game.

Stern: It wasn't that much fun?

Guyer: It didn't make much difference because we were laughing so hard that nobody stayed in the same place, and everybody was fully aware that we were in a closer proximity to one another than we ever would find ourselves in a business setting or a social setting. And that included a very expectant mother. At that moment in time, I realized, "There is something really fun here—with a capital F."

Stern: And what year was this?

Guyer: It was probably early '65. And I developed a game called King's Footsie, where people had colors wrapped around their feet, and they tried to perform several different games on a grid on the floor with these colors attached to their feet. I took it to 3M, who had a game division at that point. Their games were mostly intellectual games. They were excellent games as a matter of fact. Games called *Acquire* and *Connect*, which still exist and are still sold. But they turned it down, quite naturally.

Stern: Was it too racy for them?

Guyer: It was totally out of their comfort zone. They had these very fancy boxes, and it was totally out of their style. They were very upscale, high priced, and excellent games. But this didn't fit that at all. They all liked it, and they tried to fit it in, but they just couldn't. They knew it was a little risky.

So the game was sitting on a table in the office of our purchasing agent when a fellow who was selling silkscreen materials tried to get our business. This fellow saw the King's Footsie game sitting on this desk and asked, "What's that?" And the purchasing agent said, "Oh, that's a game that Reyn has been trying to develop."

[The salesman] said, "I've been in the game development business. . ." Well, one thing led to another—the fellow's name was Charles Foley—and I hired him.

My father went to the bank and took out a loan to back the development of the new concept. I've forever been amazed at the courage and trust in me that led him to try this high-risk possibility of starting a small division that would try to make something out of the games where *people are the players*.

Stern: You were going to put the product out in the marketplace on your own?

Guyer: No. That was not the idea. The idea was to develop the games and take it to someone who was in that business. We didn't have in mind starting a business from scratch. So we hired Foley, and he hired a fellow by the name of Neil Rabins. We got them into an office area separate from the other group, and they and I worked out about eight games that were all played by people on mats on the floor in some form. All the way from small children's games to new games like *Twister*, which we called Pretzel.

We took them to Milton Bradley, and a fellow by the name of Mel Taft liked what he saw in the Pretzel concept. He said, "Let's see if we can get some interest here in the company," and he did. And they began to work it up and develop some support for it. They had a TV spot ready to go and we had arranged to find a company that would make the mats.

Stern: You found a company to do the manufacturing of the product.

Guyer: Yes, Chuck Foley did. And that settled it for Milton Bradley. They went into producing.

Stern: Did you get a patent on this or a trademark? How were you protecting yourself at this point?

Guyer: At this point in the toy game industry, there was, and still is, an inherent, built-in need for the companies to rely on outside inventors to bring them ideas. It is not requisite that there be a patent or a copyright on any of the materials. You sign off on them, and they acknowledge and describe the inventions.

There was no real fear of someone else coming in with something like it. At that point, there were no patents or anything. Several months after *Twister* was in stores, Milton Bradley applied for a patent. It noted our employees, Foley and Rabins, on the patent because they were the ones who thought of the fabulous idea of getting on the mat with the players' hands as well as the feet. The essential idea that sparked all of our hard work, i.e. "people are the players," was not a necessary claim of the patent.

Milton Bradley showed the game to some retailers. It did not get a big response. When they took it to Sears, which was the largest toy retailer at the time, they said it was too risky—too risqué actually—and turned it down. Mel Taft called me—it was just before Christmas time—and said, "I'm sorry. We have decided to pull the product. We're pulling all the spot ads. We're just not going to do it." We naturally were very disappointed. But unbeknownst to anybody else in their business, and unbeknownst to us, Mel had already paid a PR company to get the game on Johnny Carson's show, *The Tonight Show*. So Mel and the PR executive went to the show and they put it on, sure enough. Eva Gabor enticed Johnny onto the mat, and it changed everybody's mind at that very point.

Stern: Overnight, it was a success?

Guyer: Overnight. Milton Bradley had some of the product at Abercrombie & Fitch in New York, and people were lined up at the counter fifty deep the next morning. Milton Bradley realized, "Wait a minute. We may really have something here."

So they put it back in their production line, and it became the toy of the year—the game of the year.

Stern: How did that success change the work that you were doing at that point?

Guyer: Well, Foley and Rabins decided to find another individual to back them for a game development company. They chose to go off and try it on their own. I went on to make the design agency, the Reynolds Guyer Design Agency, a success. But certainly the largest part of the success was the *Twister* game. After a while, I decided that I really wanted to try to be a toy and game inventor. I said to my father, "You know what? What if I take my portion of the *Twister* contract? You take your company and the other part of the *Twister* game royalties. I'm going to try it on my own."

I took about four artists and designers, and ended up with about six coworkers. Every day we would go to the office, right in the same building that my father's business was in, and we would go to work to design toys and games.

Stern: What was the motivation that you gave these people every day to think of something? What was the process that you had set up to motivate them or to be creative?

Guyer: Motivating a group to be creative is not difficult. I believe that sharing ideas is the secret to developing new concepts, if you can create an atmosphere of honestly sharing whatever comes to your mind. I don't care how brilliant you are or how slow-thinking anybody is, if they are willing to open up, and let whatever they say or think come to the floor, and share with everybody else, even if it's the stupidest idea, and they know it's stupid when they say it, it still often leads to something.

I've seen it time and time again. When you get two, three, or four people together in a room, and you just open up and say whatever you're thinking, good things will happen. So motivating people was not difficult. Everybody enjoyed it. It's a fun business.

Stern: You were not really working alone? You were always working in part of a group?

Guyer: No. We worked as a team. As an example, there were no desks that were set apart from the other people. Our drawing boards and desks were all around in a circle. So we all knew what each other was doing.

Stern: How did you start the day? Did you give everyone a problem for a specific field to investigate?

Guyer: We would have meetings every Monday to talk about what the agenda would be for the week, what we would be trying to do. Everybody would share their project—two or three people might be working on one project—and we'd share our progress, or lack thereof. We'd talk about possibilities. Some wonderful things came out of those meetings.

Stern: When you hired these people, were you looking for any particular skill set you thought was necessary to being an inventor?

Guyer: For the most part, they were artists. They were commercial artists. Not all of them were. We had a sales director who was responsible for keeping in touch with the game companies. But even he was involved in the creative process on a daily basis.

Stern: When you were brainstorming the ideas, was there any format that you thought worked well or didn't? Did you sketch? Did you make prototypes? Did you work in 3D? Did you have any favorite materials to play with?

Guyer: I have a set response when people ask me, "I have this idea. What should I do?" The first thing—and often the only thing—I tell them is to make one. The first job of the creative person is to make a prototype so that someone who is not familiar with what you are talking about can see what it is you are thinking.

Stern: So drawing is not as good. You need a prototype.

Guyer: It's important to get it out of your head, out of your brain. I have a number of people talking about all the ideas they have that come into my office and say, "I've this idea and that idea."

And I say, "Where is it? Make one. Show the world what it is."

That's true any time you are developing some new concept. You have to be able to let the individual who wants to see the idea use it. You have to let them use it.

Stern: So you've set up this office. You had a bunch of people creating things all day, making models and prototypes. What were some of the outcomes of this effort?

Guyer: Well actually, there were some very, very unique ideas all the way from dolls to board games. There are still some board games that I'm convinced would be fabulous board games that have never made it to the market. But what was by far the largest success that came out of that group was the idea of making products out of expandable polystyrene.

Stern: Was that a new material in the marketplace at that point?

Guyer: Well, it was. People were toying around with using it for packing material. They were using it for mattress covers, but usually it was packing material. It would slice with a hot wire.

One day one of our team members was trying to make a game that was kind of a dinosaur-era game with foam rocks [made of the expandable polystyrene material]. So somebody said, "Well, let's throw these foam rocks at each other." It was a terrible game. But Will Kruse started bouncing one of these rocks over a net, and the next thing you know, we all just said, "Okay, that's it." We started cutting out balls, little foam balls, out of the foam rock material that we had. And you know where we're going with that.

Stern: Tell us!

Guyer: Well, that ultimately became the Nerf Ball.

Stern: At this point, you're developing all these ideas. What was the process to show it to companies? Was it once a month or twice a year?

Guyer: The process was that once we made the prototype of something that we thought had value, we would then make an appointment with the game and toy company, and we'd go out and see them. We developed toys and games for a lot of people, but the one that really stuck to the wall was the foam ball, which we took to Milton Bradley—who had made such a success of the *Twister* game. And they turned it down.

So then we took it over to Parker Brothers who were in Salem, Massachusetts, at that point. There was one fellow there by the name of Henry Simmons. He was the overseer of new product development. We figured that once we had developed the ball, we had to design ways to play an assortment of games with the balls.

We found a way to cut the ball with hot wire—you take a wire that is half round, rotate it in a bun of foam, and it forms a ball, which was a breakthrough for us.

We developed a whole bunch of balls of different sizes, and then we developed several games because we thought nobody's going to sell just a ball. That's not going to go anywhere. And, of course, Parker Brothers said, "What we're going to do is come out with the ball." But better do the ball than nothing at all. And the Nerf Ball became an immediate success.

Then Parker Brothers came back and said, "We want rights on everything that you do in foam." We thought about it. "Well, are we going to go with just Parker Brothers, or are we going to proliferate this throughout the toy industry?" We decided to stick with Parker Brothers.

Stern: But you didn't have any patents on this, did you?

Guyer: No. We attempted. We had a very excellent patent lawyer. We were just talking about this other day. He came very close to getting the patent on the foam ball. That would have been very unique because the ball would do what no other ball would do. Had we done one more test, I think we would have had more proof.

Stern: Why do you think the toy industry was a field that you didn't neces- sarily need patents on as compared to other industries, which have always relied upon that?

Guyer: I think it's a very practical reason. The toy-and-game people, especially in the game industry—Milton Bradley and Parker were the big game people— knew that they didn't have or could never develop a strong enough R&D department to be creative enough.

Stern: So they needed to rely upon outside sources?

Guyer: Exactly.

Stern: Right. Do you think that still exists today?

Guyer: Yes. It does. There are some companies that are more aggressively searching out inventors than others. Some of them are beginning to think that their marketing departments can develop new product—and it's a disaster.

Stern: And why do you think that is?

Guyer: I think it is because the industry is changing, and the virtual ideas are becoming the most interesting to the consumer. The apps and the game apps and Xboxes and Kinects and gaming apparatus are more exciting.

Stern: You developed the Nerf product, and then Parker Brothers decided that they wanted all your other products using the same materials. Were you involved in developing all those other ideas?

Guyer: We developed many of them, and then other inventors began coming in with ideas too.

Stern: Were you getting royalties on all these other products?

Guyer: We had an arrangement where everything that was made of foam, we would participate in.

Stern: Well, that certainly was successful, wasn't it?

Guyer: It was highly successful.

Stern: Would you say that for the next couple of years, the Nerf products kept you and your staff very busy?

Guyer: It kept us busy for quite a while.

Stern: Nerf products are still out there. Do you have an involvement with them?

Guyer: The licensing arrangement I had with the Nerf products has expired.

Stern: In looking back, the two biggest products that you seem to have done were *Twister* and Nerf. Any thoughts on how those inventions influenced society?

Guyer: Well, yes. I think the overriding gift, if you will, to mankind and society is fun, F-U-N. And if we were to be specific about each product, I would say that each of those products were successful because they broke a rule. The rule that *Twister* broke was the assumed rule or moral rule that you really shouldn't, in a social setting, be so close to another individual for a prolonged period of time if you are not dancing. It broke the rule that people really shouldn't be that close to each other. And the Nerf Ball broke a rule that you shouldn't throw balls in the house.

Stern: Certainly all our mothers told us that. In looking back at this company that developed ideas, did you have any really big failures that you thought were going to be a big hit, but that didn't work out?

Guyer: Ninety-five percent of them didn't work out. There are a lot of wonderful ideas that I still think are excellent ideas but have not yet found their market. They might some day.

Stern: You don't think you could predict which will succeed and which will fail?

Guyer: You know, I think just plain old good fortune plays an enormous part in successes. Yes, you have to be trying to do it. I think that's important. And I think it's important to stick to it and not give up on it. After all is said and done, if you are persistent and you make models that are good models, and you believe in your product—you still have to get lucky.

Stern: What have you've learned after doing all this?

Guyer: We were pretty much cutting fresh ice. We were running roads that a lot of people hadn't been down. So the contracts we were writing with the lawyers that I worked with were negotiated with the toy-and-game companies on a very fresh and new basis. So we were kind of making things up as we went along.

Stern: Do you think they appreciated that as well?

Guyer: Well, I think in some instances we wrote terrific contracts. I think they somehow wished today that they had had more experience in writing contracts.

Stern: Do you think that they paid you too much?

Guyer: No. I don't think they think they paid too much. They know that if it's a success, I too will succeed.

Stern: Do you have any professional heroes or mentors? When you were doing all this work, were there people that you admired?

Guyer: Well, the man that I really think was a pioneer was a man in Milton Bradley by the name of Mel Taft who had the courage to stick to his guns and try to prove to his company that the *Twister* game was a viable concept. Mel Taft is still alive and well, living outside of Springfield, Massachusetts, and he's a wonderful man. And yes, if I would say if I had a mentor, it would be Mel.

Stern: During the process of doing all this work, were there other inventions or inventors in the toy biz that you admired?

Guyer: Well, I admired the success that a group out of Chicago—oh I forget his name. He was a very unusual man, but I admired his team's products. You know, there are so many really creative toy-and-game designers. I really haven't spent a lot of time getting to know a lot of the other toy-and-game designers.

Mainly because I don't consider myself only a toy-and-game inventor. I do other things.

Stern: What other fields are you working in now?

Guyer: We have an education company by the name of Winsor Learning that is one of the leading companies for remediating children who are behind in their reading skills. Reading and math. This company has gone on to help schools deal with children with behavioral problems. It's quite an exciting company.

Stern: Why did you get involved in that field?

Guyer: My wife Mary and I, and all five of our children have a learning disability. All of our children have been remediated for that. The lady who had been remediating them, Arlene Sonday, who was one of the leaders in the country in the field, came to us and asked, "How do I get this across to more people rather than just doing it one-on-one?" She and my daughter Cynthia worked together for over a year to develop a step-by-step program that anyone can use. Anyone who is a helper can help a learner with this program. That was unique. Once that model was made, we had a whole different, special product that no one else had.

Stern: Is this the focus of your work right now?

Guyer: No. The focus of my work is still developing new toy-and-game concepts. We are working on several virtual games right now that appear to have great potential. We have a new lawn game called *KingsCourt*, which I developed because I don't like croquet. I find croquet quite boring. *KingsCourt* pits two teams in anybody's backyard. It pits two teams against each other trying to knock over the opponent's pins or capture the opponent's ball. Much more exciting.

Stern: Has doing *Twister* and Nerf products over the years opened doors for you that typically wouldn't be open for an inventor?

Guyer: The fact that we've had three or four good, big successes, and two that are exceptional—the Nerf and *Twister*—does give us an opportunity to talk to people in the industry. That's true. But it doesn't necessarily carry over into other industries. My daughter Ree and I started a company in Nashville called Wrensong. It's a very successful music company now. We've had two country music songs of the year and numerous number-one hits with major recording artists. But it took a long time to get that up and going. There's no carry over between industries.

Stern: Is there any particular advice that you would give to an inventor who is just starting out?

Guyer: It would be "Make one." If you think you have a concept, make one. Make the prototype. Do the model as best you can so that someone else can understand it and do it.

And the other would be work with a team, because it isn't until you get a group together that really exciting things happen. If you get a group that is willing to share their ideas, that's when the real creativity begins. I really don't think that there have been many inventions throughout the years that have been attributable to one person, solely one person.

Stern: Do you have a favorite product unrelated to toys?

Guyer: No.

Stern: What about a favorite technology that you use today?

Guyer: I think right now the opportunities lie in the development of the virtual products.

Stern: Is that the future of the toy industry or just in the design field?

Guyer: Well, I don't know if it's the future, but it certainly is a favored product of the consumer right now. Whether that'll last, who knows? I know they're popular now.

Stern: When you're at a dinner party or meeting someone new, what do you tell them what you do?

Guyer: I do have a problem in filling out my visa when leaving the country or going to another country as to try to describe what I do.

Stern: So what are some of the things that you put in?

Guyer: Well, I tell them I'm an inventor, artist, and writer. I guess I didn't mention that the latest product that I've been working on is a musical called *Stained Glass*. We had a fledgling musical up in Connecticut called *Spirits* a few years ago. And now my cowriter, Jeff Harrington, and I have developed one called *Stained Glass*, and we rehearsed it in New York with a New York cast and made a video of it in New York recently. We're going to take video shortly and begin sending it out to small theaters. We also put it up in a small theater near my home here in Florida, and it seems to be very successful.

Stern: What do you do for fun or distraction? Because you have so many different things you're working on, but what do you do that you don't consider work?

Guyer: I've seldom ever made a separation between my work and my play.

Stern: Do you plan to retire?

Guyer: I wouldn't know what that would mean. I wouldn't be having this much fun.

Stern: Anything final that you want to say to the people reading this?

Guyer: I do. I really believe that the ideas that have been put before me, that I've been a part of, are my gifts. That sounds hokey, but I really believe that I have a gift, and I see it in others, and that I have been given these gifts, these concepts. I guess I feel responsible for doing the best I can to bring them to the world. That's my job. If I don't succeed in bringing it to the world, either it's not supposed to come to the world, or I didn't do a very good job of getting it there. I think sometimes that's the case. I just haven't really done the proper job of getting the product out to people because I know that I have a gift of understanding what is fun. It's our job to do the best we can with them.

Bernhard van Lengerich

Chief Science Officer and Vice President for Technology Strategy, General Mills

Dr. Bernhard van Lengerich is chief science officer and vice president for technology strategy at General Mills, Inc., in Minneapolis, Minnesota. Dr. van Lengerich studied food and biotechnology at the Technical University of Berlin, Germany, and completed his dissertation on systems analysis and food extrusion in 1984. His work experience includes employments at Unilever Germany, RJR Nabisco in New Jersey, and Buhler AG in Switzerland. In 1994, Dr. van Lengerich joined General Mills, Inc.

During the course of his career, he led the development and implementation of many enabling technologies that resulted in numerous new and healthful food products. Dr. van Lengerich has been issued more than 80 US and international patents for his inventions. He has published over 50 international publications on systems analysis and food extrusion. Dr. van Lengerich serves as honorarium professor at the Technical University of Berlin and as an adjunct professor at the University of Minnesota in Minneapolis. He was named a Fellow of the Institute of Food Technologists in 2011, and he recently joined the European Technology Platform "Food for Life" executive board.

Brett Stern: If you can give some background—where you were born, your education, and your field of study.

Bernhard van Lengerich: I was born in Germany, the oldest son of a family of bakers. The first part of my education was as a three-year apprenticeship as a baker and a two-year apprenticeship as a "konditor," which is the equivalent of a confectioner or pastry cook. I learned how to make bread, puff pastries, sweet goods, pralines, and truffles, among other things, at a very young age. According to the tradition of the time in Germany, the oldest son leads the family business after practicing an artisan craft for at least five years following an apprenticeship. The passing of my father, who was the head of the business, at an early age, prevented me from pursuing this traditional route, so I was sort of stuck and needed to find an alternate path forward.

What followed were three years working as a baker and pastry chef while attending night school in order to prepare for college. Early on, I learned that if you set your mind to something, you can do it. In college, I studied food technology with a focus on baking and cereal science. I didn't want to stray too far from what I really liked—bread, baking, and food.

My first job as a food technologist was at a global company, based in Bremen in Germany, which produced ingredients for the food industry, primarily bakers. I worked in their application and development department and absolutely enjoyed the experience. It gave me the opportunity to develop food products that were exciting to consumers and were the ingredients bakers in Germany relied on.

After a few years, even though I essentially had a fantastic job already, I decided to continue my education at the Technical University of Berlin to work toward an MS and PhD in Food and Biotechnology. I wanted to ensure I had the level of education that I'd not be limited in what I could do. I conducted my thesis in the field of Cooking Extrusion of Starches and Flours, which was in the infancy of its study at the time. A cooking extruder is essentially a machine that cooks grains and presses them through an opening. It has wide use in the food industry and is used for making breakfast cereals, pre-cooked flours for sauces, snacks, flatbread, and many other foods that are based on grains or flour. While some companies already had extruders at that time, many didn't yet know how best use them, which presented a great opportunity for my studies.

Because of my PhD work with extruders and food ingredients, and food extrusion technology was advancing fast in the US at that time, I was offered a job in the US by a company that built the machines that I worked on. I accepted it with great excitement and moved with my family to the US with the belief that we would only be in the US for two or three years before moving back to Germany. That was over twenty-five years ago, and here we are, still in the US.

Stern: From these machines, what type of products were you developing?

van Lengerich: As mentioned, extruders are able to produce many different kinds of foods: breakfast cereals, snacks, pre-cooked flours for baby food, pasta, or even pet food. They are very universal machines that compress, cook, and shape grain-based food. I was privileged to have been among the first generation of students to work with this then-new technology of "twin screw extruders" from a scientific and engineering perspective. During my first years after completing my PhD, I was able to meet and work with many engineers and scientists from various companies in support of their development projects. Quite frankly, I believe one of the major skill sets for an inventor or an innovator or a creative person is to have many dots to connect, which is why this experience was so formative for me.

Stern: It sounds like you almost had a clean slate of what could be done.

van Lengerich: While it was a very rich experience and eye-opening for me to see and work on the variety of needs and challenges that existed, there were usually time constraints under which projects needed to be completed, which was great learning. Within short time frames I was asked to help teams test food formulations, process technology and machines in order to make decisions on very sizable investments. Most of this work took place in the laboratories and pilot plants of big and small companies across the food industry in the US and abroad, many times with the purpose of developing new foods that could only be produced using an extruder. It was a very rich experience with a lot of great exposure to industry needs in the real world, and filled with opportunities to work with different teams in very different environments.

After several years in this position, I was given the opportunity to join a large food company in the US, and my team and I were given the challenge to develop new products and strategically important technologies for the company. It was during this time that many of my early patents were issued. In fact, one key invention, a method to produce cookies using an extruder, was covered initially by thirty-six patent applications.

Six years later, my family and I followed our original plans to return to Europe and I accepted an opportunity to lead a global R&D organization at a Swiss company, developing technologies and equipment for the food industry. This position, again, offered rich exposure to and collaborating with leading teams to solve complex and new technical challenges that required new thinking to find novel solutions—it was a great experience and different than any of the ones I'd seen before.

In 1994, we brought our family back to the US when I joined General Mills in Minneapolis with a focus on technology development. Besides my previous experience with extrusion technology, the company also was interested in my diverse set of skills and I found that my previously built network has made it easier doing my job. I feel strongly that having and maintaining connections to people in your professional network is an essential part of both the innovation and invention process.

Stern: Can you talk about your ideation process of how you go from defining a problem to finding a solution? Do you have a particular method for brainstorming or anything?

van Lengerich: A critical aspect of coming up with an idea is to clearly understand the need for developing a solution.

Stern: Are you developing the needs? Or do the needs come from the marketing department or the consumer?

van Lengerich: Generally, the needs come from the consumer. In the environment in which I've spent most of my career working, the needs always come from the consumer, whether directly or indirectly. However, those needs will need to be translated into technical opportunities.

For example, we know that many food ingredients that we may want to use have a health benefit, but they may not be palatable whether due to a bitter taste or some other reason. Even though consumers would like to have these ingredients in their food for a health benefit, we can't compromise its good taste. So, wouldn't it be nice if we could find a technology that changes or encapsulate these ingredients into a form that they become invisible to the palate? In other words, we could deliver the health benefit without the bad taste. This would be an example of a technology that benefits the consumer and delivers health without compromising taste. We have a series of patents for technologies to encapsulate nutritionally beneficial ingredients such as omega-3 fatty acids or vitamins to make them palatable for food or to protect their activity such as probiotic cultures.

Stern: Are you working on these projects as part of a team with people with different backgrounds, or are you by yourself in the lab?

van Lengerich: I work only with teams and people with a variety of different backgrounds—engineers, scientists, biologists, nutritional scientists, etc. In order for an initiative to be significantly impactful, each person on a team is critical and one person alone won't be able to do it. The initial technical idea or an original concept might be one person's contribution, but it will need to be reduced to practice in the lab, scaled up, utility-tested, and vetted. A fairly big team of people is involved in any new product or technology for these steps, from the conception of the idea all the way until a product ends up on the grocery shelf.

Teamwork and diverse thinking is critical for success throughout this collaborative and sometimes complex invention process. I encourage people on my teams to leverage and share their diverse backgrounds in order to help us reach a better solution.

Stern: Do you have any thoughts on what makes a good team and how to manage that team?

van Lengerich: As mentioned earlier, one aspect of an innovative team is to have every team member clearly understand the goal and solutions for which the team is working. The team leader should empower the team members to leverage each individual's full potential and remove hurdles that may hinder the team to move forward. At the beginning of an ideation process our teams alternate between diversion and conversion, and then focus in on executing the solution with the highest likelihood of success.

Stern: Food is a unique area for innovation. Would you say that a lot of your development process is "hands-on" experimental right from the beginning?

van Lengerich: Yes, absolutely. Innovation does not happen in an office. It may be possible to conceive an idea during a discussion, but usually only if there has been some prior experimentation. Most ideas are created in the pilot plants, in the laboratory, and in the production environment.

Additionally, there is usually a high degree of scientific and practical maturity, and people on the team have good judgment. And while curiosity is important, they quickly know and are able to exclude many things that won't work. In our experience, one good practice is to do "the last experiment first," which is essentially a set of experiment conducted under conditions chosen collectively by all team members to be most likely to be successful considering the mature technical judgment by the team. This allows us to go from a very high number of possible experiments, sometimes several hundreds, to a very manageable number of maybe fifty or ten. With a much smaller set of systematically designed experiments, we can quickly learn and iterate the process in the direction of what's working.

Stern: You just were talking about judgment and maturity. Do you think there are any personal skill sets that you need to be a good inventor?

van Lengerich: Yes, I believe that it is a combination of a number of factors: ability to connect many dots, creativity, judgment.

Stern: What makes someone creative?

van Lengerich: In our world, one description of creativity would be the ability to create and articulate a reasonably plausible concept that would be outside of an existing paradigm.

Stern: You put all these attributes together and you have this team that is going full force. How do you control them to make sure they get to the end and that they have the right motivation?

van Lengerich: Teams are motivated by interdependency, trust and the realization that at the end of the road is a successful product in the market with a clear consumer benefit.

Stern: You were just talking about succeeding. Is it only a success when it's financial success? What if it's a technical success but doesn't sell well? How do you rate that?

van Lengerich: This is a good point. Even without commercial success, the learnings of a technical success are of high value. At General Mills, we conduct an annual technical conference, where people in our technical community share with each other technical accomplishments and successes they have achieved over the last twelve months independent of commercial success.

We found that this event is an extremely valuable venue and forum to inspire new ideas, new inventions, and new processes for a variety of product platforms. For example, someone working on a new yogurt product may actually face a similar technical challenge as someone working with Lucky Charms marshmallows—however, they may not work together in their daily lives. The technical conference provides them with an opportunity to connect and exchange ideas that can ultimately benefit the company.

Stern: I realize that you're more in management now, but where do you seek inspiration or possible solutions outside the lab?

van Lengerich: Today I find myself curious about what happens in the general sciences. I wish the day had more hours so I could read everything I wanted to. I find it interesting to read and learn about all kinds of advances in various technical and scientific fields in life sciences, engineering and technology, not necessarily related to food.

Stern: Along the way, have you had mentors. What do you feel that the role of a mentor should be?

van Lengerich: Trust is a key ingredient between a mentor and a mentee. A good mentor certainly provides honest help and the best advice possible, points out areas of concern, if appropriate, and can also act as a compassionate partner to listen and provide thoughts for advancing ideas. It's important to have a network of people who you trust to bounce ideas off, or just to get a different perspective.

Stern: Do you have any particular advice that you would offer an inventor coming into the marketplace?

van Lengerich: If you have a novel idea that satisfies a need and is usable, keep at it. Don't give up. If it does not exist yet, and it has utility, I'd suggest meeting with a patent attorney to describe and articulate the idea. If you're not able to articulate or demonstrate it, then it's of little use. Being able to communicate the solution to others in some shape or form is critical.

Stern: It's interesting you say that, because most inventors are not very talkative people. Are there some exercises or techniques that you would suggest to inventors to help them communicate their ideas better?

van Lengerich: Sometimes it is difficult to communicate an invention in words. When discussing an invention with a patent attorney who has a technical background by training, it will become easier. Another way to effectively communicate is by showing a working prototype apparatus, a product, or any other real-world demonstration of the invention.

Stern: With the work that you're doing at General Mills, is there a future generation of ideas of where you see food science going?

van Lengerich: Yes. Looking at the year 2050, the world will have approximately nine-and-a-half billion people, and I believe that securing food and water for all will be the most relevant priority we face. Today, key challenges and opportunities for innovations related to food security are to reduce harvest losses, particularly in the developing countries and continents that are approaching very high loss levels of sixty to ninety percent. There are many opportunities to invent using today's knowledge base, simple devices, processes, and technology solutions for minimizing harvest losses but also to teach and transfer production technologies to manufacture safe and nutritious food for people in need.

Stern: With the products that you have worked on in the food industry, do you have a favorite food item that you just love to eat?

van Lengerich: Yes, Cheerios! They're made from oats and are high in soluble fiber. I believe they are good for your cardiovascular system and keep my cholesterol in check, and are a good food. I love to eat Cheerios.

Stern: Do you ever look to the general public for new food ideas?

van Lengerich: Absolutely. If somebody has a good technology or solution, they can submit their ideas to General Mills. We have a process in place on how to evaluate and potentially channel all these good technologies into the right businesses. We call it our "connected innovation" concept or the General Mills Worldwide Innovation Network [G-WIN]. It is effective and allows us to develop better and bigger ideas, and be faster to market. We learned that it is critical that our needs are very well-articulated, then we publish those on our G-WIN web site and invite inventors to submit proposals.

Stern: When you're at a dinner party, what do you tell people that you do?

van Lengerich: You know, I really don't go to many dinner parties.

Stern: Do you have a next field of study that you're working on right now that you could talk about?

van Lengerich: One set of ideas we are working on is to enable the addition of healthful components to food, and an even further reduction of sugar and sodium to make our food even more nutritious and healthful, without negative taste implications while, of course, keeping it affordable to the consumer.

Stern: When you're not doing your work, what do you do for fun?

van Lengerich: I love to spend time with my wife and our three children. I enjoy being outside, running, and skiing. I like to travel, experience different countries and cultures. I like to eat different types of food.

Stern: Do you plan to retire? And what will you do?

van Lengerich: I have no plans to retire and I love what I do now.

Stern: Any final words of advice that you want to offer to an inventor?

van Lengerich: See the glass half-full. Don't give up. Successful inventions can be compared to looking at an iceberg: success is what you see above the water, but you don't see the unsuccessful experiments underneath, without which there would be no success.

Curt Croley

Senior Director

Shane MacGregor

Director, Advanced Concepts and User Interactions

Graham Marshall

Director, Global Design and Research
Innovation and Design, Motorola Solutions

Curt Croley is senior director of the Innovation and Design group at Motorola Solutions. He has a degree in industrial design from Kent State and has taught design at the University of Cincinnati and the Cleveland Institute of Art. He has been issued 45 patents in connection with his design work for Motorola Solutions, and 12 more are pending.

Shane MacGregor is director of the Advanced Concepts and User Interactions team of the Innovation and Design group at Motorola Solutions. He graduated from Syracuse University with a degree in industrial design. Since then he has held various positions *within the Motorola Solutions design group in the areas of user interface design, industrial design, and advanced concepts.*

Graham Marshall is the director of the Global Design Research team of the Innovation and Design group at Motorola Solutions. He has led design teams at NCR, AT&T, Nortel, Esion Networks, and Kaleidoscope. He has a degree in industrial design from Carleton University and subsequently taught product design strategy and innovation there. He has received 18 design awards and been issued 13 patents.

Brett Stern: I hate to tell you this, but I didn't have the record-button on. Sometimes there is too much technology. If we can, let's start again with everyone's introduction.

Curt Croley: I am Curt Croley. I am the senior director of the Innovation and Design group here at Motorola Solutions. I have a degree in industrial design and have been practicing for the last twenty-seven years with a focus on handheld products and high-tech devices.

Shane MacGregor: I am Shane McGregor. I am the director of Advanced Concepts and User Interactions team in the Innovation and Design group. I have an industrial design background primarily focusing on technology devices, looking at both hardware and user interaction.

Graham Marshall: I am Graham Marshall. I am the director of Global Design Research. I also have industrial design degree and have worked in a number of different corporations—including AT&T, NCR, and Nortel—managing different groups doing industrial design, graphic UI, and research in human factors.

Stern: Can anybody give me a definition of what industrial design is and how it is different than engineering?

Croley: The practice of industrial design is to synthesize marketing, engineering, and customer research into product visions to solve our customers' needs in a unique way never seen before. Then those design visions get translated to the engineers for additional technology development and detailing.

Our Innovation and Design group is an umbrella for four disciplines: customer research, industrial design, human factors engineering, and user interaction and experience design. This allows us to mix it up across all four disciplines very quickly to come up with what we feel are the right solutions for the company. Now in the context of that, Graham take it away from a customer research perspective.

Marshall: From a customer research perspective, we want to innovate around our customers' business processes rather than around our product or device. So, our innovations flow from doing a lot of observational research, informed by intuitive insight about what the customer really wants to be doing.

Croley: And it is also important to know that when you ask customers what they want in a future device, what they say they need doesn't necessarily jive with what you observe they need in the workplace. They are looking at solving their immediate problems, so their orientation is very short-term.

Stern: Are the products that you are working on consumer-based or business-based?

Croley: Completely business-based. We do no consumer devices. The devices we are working on are changing the way people shop, the way people have packages delivered, and the way businesses operate in many respects. We are envisioning how to make the checkout line faster or maybe eliminate it completely. Have you ever wondered how you could track a package anywhere in the world at a moment's notice? That is our technology. Although what we are doing has a dramatic effect on customers around the world, we are not designing devices for consumers.

Stern: What is "innovation" and how is it different from "invention"?

Croley: Innovation is coupled, in my opinion, directly to insight. You put discrete ideas together in a unique way to create an insight. That insight can be translated to an innovation that could take the form of a new product or a new business. Invention is a process of rigorously applying science to solve a problem. I view insight and innovation as separate from invention. What do you think, Shane?

MacGregor: I think that innovation comes from the insight that comes from reflecting on observations you make in the field.

Marshall: Innovation is related to the big picture of what we are trying to do and how we are trying to change things. Invention is getting down to the nuts and bolts of solving specific problems.

Stern: Does commercialization drive innovation?

Croley: Innovation can come from technology, market need, or user need.

Marshall: Constraints such as existing technology drive a lot of invention, insight, and innovation. The more constraints you have, the more creative you have to be to get around them.

Stern: So constraints are good?

Croley: Not always. But having to work within such constraints as the laws of physics and the realities of business timing can help drive you to a better solution.

MacGregor: Let's face it: it is very rare to have a clean sheet of paper in front of you. There are always constraints. If people had four arms, we could design devices with two more degrees of freedom. But obviously we have to design within biomechanical constraints.

Stern: You talked before about researching the needs of your customers. How do you do that?

Marshall: There are three ways that we get information from our customer base: surveys, focus groups, and site visits. We do a lot of site visits. The good thing about focusing on enterprise and government is you can target the different user experiences in controlled settings.

Last year, for example, we have sent six researchers out to two hundred sixty customers in Europe, South America, and Asia—so in effect we observed each customer for a full week. We are constantly out in the field conducting observational research and on-site interviews with our customers.

Stern: How do you document your observations?

Marshall: If allowed, we use video and photography to share our observations with our design teams in cross-team brainstorming sessions.

Stern: Could you talk about the brainstorming and what your process is?

Marshall: As soon as we feel we have enough information and perspective on the customer's procedures and sites, we bring in a multidisciplinary team of designers and engineers to do first-level analysis and categorize where the problems and opportunities are.

Stern: How many people are in those teams?

Croley: I prefer ten, or twelve at the most. They can get too big—thirty people sometimes.

Stern: In these brainstorming sessions, you're throwing out ideas or words or sketches and you're not filtering anything at the beginning. Could you talk about how you go about filtering your collective thoughts?

Croley: That is a good question. It is usually by finding logical groupings and/or categorizing or what we can do now or what we can do later on. It is also interconnected to the problem that we are trying to solve. You have to figure out as you go along, "How can we best structure this based on the problem we are solving right now?"

Marshall: We are also alert to wild-card moments in these brainstorming sessions, when someone has an insight or suggestions that may not solve the problem that was put in front of everybody, but that has the potential to take the business in a completely different direction. You put those off to the side for further investigation.

Stern: It sounds like you are continuously balancing short- and long-term opportunities. When you are doing these brainstorming sessions, are you thinking of the next product that is going to come out?

Croley: Yes, our innovation design team has a dual role, reporting directly to the CTO. The immediate problems that we set in our brainstorming sessions are in the one- to two-year range. But we also have a second category of projection, called Horizon II, looking farther out to five years or eight years.

We employ the same process, but we apply a more intuitive and holistic understanding of where we believe the businesses are going.

Stern: Is there a Horizon III?

Croley: Yes, Horizon III involves the invention of enabling technology. So we attach people, perhaps not as part of our team but as part of a broader research effort, to go off and invent it.

MacGregor: The role of the innovation team is to bridge the gap from today to tomorrow as innovators within the company. To do that successfully we have to refine our ability to solve today's problems as well as pitch where we want to be in the future, and create the threads between the two. It is through that pitch that we start to break things down into smaller projects that can be worked on individually. Then the goal is to bring those back together as a complete solution.

Croley: For Horizon II, it's not good enough to say, "Hey, I've got a vague notion that there is potentially some opportunity here five or eight years out." We really have to articulate a vision around which people can rally and say, "Okay, I get it." Then we say, "Here are the technologies we think that we might want to employ." So our team paints that broader vision with storyboards or more sketch-based scenarios.

Marshall: Sometimes we produce short videos.

Stern: So, after brainstorming and whittling down the problems to specific projects, how do you decide who is going to work on the projects? How do you decide who is going to champion them, and how do you put the multidisciplinary teams together?

MacGregor: If we put our brainstorming session together properly, we will be going into that session with three or four problems that we need to solve already identified. Our ideas are focused in those areas. We always get outlying ideas or ones that don't fit within those buckets that we definitely want to capture. Once we capture those ideas, we focus down into the areas that you were there to solve. From there the team that initiated the brainstorming session takes the focused ideas and goes off to solve the set problems, while the farther-reaching ideas are taken up by a group that has been identified to look at the next horizons.

Stern: It sounds like you are continuously getting customer input as well.

Croley: Yes, we don't just talk to customers once and then check the box and go away. We are talking to customers and cycling back with them continuously throughout our entire innovation process. We want them to feel they are part of the development process. The biggest compliment that can be paid to our team is if our key customers at the end of the process say, "I designed that."

Stern: So they are taking ownership of it.

Croley: Yes, because they really feel like it is their solution and it is their insights that are being manifested.

MacGregor: It's not like in the mass-market consumer world where you are selling one device to one individual. Some of our customers are very, very large and they are buying our products in the thousands. So engagement with customers throughout the development is really important.

Stern: The inventing process is somewhat of a singular or individual event. But it sounds as though you are always doing everything as part of teams working on numerous ideas and directions. How do you strike a compromise between the individual and team approaches?

Croley: I don't find a conflict here. When two, or three, or four people work together collaboratively and feed off each other's ideas and energy, the sum is greater than the parts. A healthy discussion of ideas is definitely a group sport for us.

MacGregor: We have to leave our egos at the door and become part of a holistic team. Everybody has his or her strengths. It pays to know what our individual strengths are and how we are going to best contribute to the idea or the topic in collaboration with people who have complementary skills.

Stern: What skill sets do you think a person needs to be innovative or inventive? What are those personality traits?

Croley: That's a really good question. I don't know if we've ever talked about that.

Marshall: From my point of view, being a good observer and a good listener on the research side is really critical for us.

Croley: Yes, those traits are critical for inter-team dynamics. We need people who are able to listen and synthesize other people's ideas and then add their own thoughts. As Shane said, egos must be left at the door. If people just proffer their ideas and their ideas only and don't want to listen to anything else, that probably isn't going to work within our organization. What we see in a properly functioning team is everybody getting excited as we start building off each other's ideas to create an outcome that no one person could have envisioned.

Stern: What tools do you guys use to ideate and brainstorm? Do you sketch? Do you prototype or do computer modeling?

Croley: All the above. Our team is multidisciplinary with a range of tools at its disposal, including Alias AutoStudio, Autodesk for high-end free-form 3D surface modeling, and all sorts of Photoshop and Illustrator tools.

Marshall: At a fundamental level, individuals select and hone whatever tools they need to convey an idea. A storyboard or sketch might best convey how something interacts within an environment. Prototyping might best convey an interaction with an object: the good old-fashioned method of bolting a couple things together to get something physical. The quicker we can get to something tangible, the better off we are.

Croley: We employ a lot of 3D rapid prototyping and a lot of rapid visualization. The important point here is that we are not an island into ourselves. Our innovations, or ideations, or brainstorming all have to be communicated to other people to help align the organization around a particular vision. We need to speak the language of the various people involved in realizing the vision. So, if we are trying to convey an engineering-oriented idea of a physical object, we will articulate the idea using the engineering tool sets that our engineers are versed in. If it is an interaction design with a very sophisticated workflow that we want to convey, we will make a flash prototype or some paper-based visuals that demonstrate the intent. Ultimately, if we are designing a physical object, we will sculpt it three-dimensionally in a computer and manufacture a rapid prototype of that.

Stern: Where do you each individually seek inspiration from?

Croley: That is always a great topic. It is something that we try and invest a lot of time in. So, there are two parts to this question. The first part is, "How do we derive inspiration as a group?" We structure a lot of activities that allow the group to come together as a team and feel a group dynamic. It can be stress-relieving events like have a drink after work. When we get team members together into a social setting, a lot of ideating happens on bar napkins.

But then the second part to that is how the people derive inspiration individually. From that perspective, it's particularly inspiring to have a multicultural team where everybody is bringing their different viewpoints to bear. People get excited when they see stuff from somebody else's private world. Insights can even come from just sharing anecdotes about what someone did on the weekend.

MacGregor: The core of inspiration is observing the world around you. In the world are a lot of very creative people who are doing a lot of really good things. We also get inspiration from watching the environment of use—what people are actually doing. When we are out there in the world, especially in some of our more extreme environments such as in public safety, we watch people using our devices and we are inspired to make their job better by what we see they are trying to accomplish.

Croley: One of the most humbling things for us is hearing some of our public safety customers tell us they would give up their guns before they give up their Motorola radios, because their Motorola radios mean life to them. The chief of the Boston Police Department told us that our radios have saved more lives than the Kevlar vest he was wearing. When we hear comments like that, we realize we are doing something that is really helping people come home at night to their loved ones, and we sense the gravity and the meaning of our work.

Stern: Most people work at their job nine to five. It sounds like this process you go through is more of a 24/7 commitment, continuously being an observer?

Croley: It is both a blessing and a curse. There is not a person here who works the nine-to-five gig. That moment of inspiration happens anytime day or night. We all have notepads on our nightstands.

Marshall: I just wanted to add to what inspires me is when I see a product that has solved a problem I didn't realize was there, and we look at the product and say, "Oh, of course! That's the way it should work. Why didn't we do this before?" I just love seeing something really new that solved some unusual problem, even including our own in our backroom managing devices. It was such an obvious answer—such a brilliantly simple answer.

Stern: Why is simplicity so hard?

Croley: The simple answers are not necessarily easy. People oftentimes are looking for a much harder answer or solution.

Stern: Can you talk about failure and what your motivation is to get over those mistakes?

Croley: It depends on your definition of failure. I guess I don't have any failures. I just have a whole lot of ideas that didn't work very well, or I have another one that worked better than the last one.

Marshall: We feel everything is a process. When we look multi-generationally across the whole host of our ideas, projects, and products, we see that each one of them solves customer problems in a compounded way over the last generation. I don't view any of those previous generations as failures.

As our insights and our understanding of our customers become deeper, we recognize that we can always go deeper on the solutions we are providing. I think there is not a person here that would say, "Ah, success! We have hit the ultimate solution and there is no room for improvement." failures

Marshall: I think pushing past failure and success is a personality trait. Innovators are people who are continually pushing their limits. They know that there is always something more that they can know. Knowledge is an ever-receding goal that we are always going after. It is that striving to find a better way to do things that I think inspires us to push to that next level. failures

Stern: Are there other personality traits that you see in yourselves that make you creative people?

MacGregor: Not being happy with status quo is one—the recognition that there is always a richer solution just on the horizon.

Stern: Along the way, you have had mentors. Could you talk about the benefits of having had mentors, how you found them, and if you are now mentors to others?

Croley: Each one of the three of us is a mentor to others within and outside our teams. Along the way, there have been people who have strongly influenced my life, even to the point of why I got into industrial design. My high school engineering teacher gave me a book on ID by Sid Mead. Right away, I was sold. Other people throughout my career have provided guidance as to how to enrich myself personally through my profession and have really helped me become who I am today. I am very appreciative of that.

MacGregor: My mentors in my career have taught me the invaluable lesson that is that it is one thing to have the nugget of an idea, but it is another thing to build that idea, convey it to other people in the organization, and bring it into reality.

Stern: All of you are industrial designers. Would you also call yourself inventors? Is there a difference?

Croley: I think all of us would call ourselves inventors.

Stern: Would you say most industrial designers would call themselves inventors?

Croley: I think the design thinking process enables a lot of industrial designers to think more broadly about problem solving, but I believe that innovation goes beyond the problem in front of you to a vision of future possibilities.

Stern: Any thoughts on the future or next generation of where your technology is going?

Croley: We are in a golden age of technology right now. It is just absolutely amazing. Motorola is on the forefront of harnessing this technology to deliver new goods and services to our customers. We are adapting and refining this technology so that it is responsive to the context of the user in the workplace, serving up the information the user requires without the user having to ask for it. We see this kind of powerful smart technology as the future.

Stern: Do you think our daily environment is changing more than during the industrial revolution in how people live and act?

Croley: I think we are going through an analogously profound technological revolution.

Marshall: There is so much computing power in our hands right now that we can rethink a lot of things.

Stern: Any advice on how to become a creative and inventive person that you could offer to would-be inventors, students, and junior designers?

Croley: My guidance would be keep an open mind and don't be too in love with your own idea. For me, the notion of innovation is looking at all of these different things happening around you and synthesizing them into a unique insight. If you go in to the process thinking you have the answer, you lose the opportunity to recognize even richer solutions.

Marshall: Perseverance as well.

MacGregor: Yes. The idea might not be perfect when you first start, but if you continue working at it through a process, you will get to that innovative idea. Also, be honest with who you are and what your strengths are. Don't be afraid to go find or talk to people around you to figure out how you make your idea better. Communication is one of the keys.

Croley: Yes, you have to be dogged in your conviction that you want to solve this problem, or you want to innovate in this area. Don't let anything make you waver from that path.

Stern: Do you guys work on multiple projects at a time and is there an advantage to that?

Croley: Yes, our teams are working on many separate projects simultaneously. It does create a din of discussions that sometimes is hard to separate, but I do find that working on them simultaneously fosters synergies.

MacGregor: Inventors and innovators usually have very busy minds. To try to keep ourselves focused on one thing becomes a challenge. We try to make sure that we have outlets for multiple tracks of thought, whether they are personal projects that we are doing outside of work or multiple internal projects that we are working on. This multitask approach keeps the ideas flowing.

Stern: When you guys are at dinner parties, what do you tell people you do?

Croley: If you have a dinner party with our team, invariably the whiteboard comes out at some point. But whiteboarding is frowned on in mixed company, so we limit ourselves to talking about the fantastical experiences we are having in customer research and design brainstorming. Sometimes we get blank stares, or people just say, "That's nice."

Stern: So, what do you guys do for fun when you are not designing?

Croley: Designing happens all the time and it's fun.

MacGregor: You don't turn it off. Usually our outside fun involves the exploration or creation of something.

Croley: By the way, we find that big goofy dogs help as well.

Marshall: Yes, big goofy dogs.

Stern: Do you guys plan to retire, and what would you do?

Marshall: I'll stay in the arts. I've done painting and sculpture in the past, and that's definitely what I'll get back to.

Croley: My ideal retirement lifestyle would be having an automotive design studio in the basement, decked out in such a way that I can host innovation sessions for young designers to come and hang out in a beach-like setting for a couple of weeks at a time and innovate. Yeah, mentoring younger designers in a beach-like setting …

Marshall: Wait! I am going to change my idea to match that one.

MacGregor: I have no plan to retire anytime soon.

Stern: Any final thoughts?

Marshall: All that we do as inventors and designers falls into three categories: understanding the problem through observation, solving the problem, and communicating the solution. Communication is just as important as the other two. We touched on it in reference to the perseverance and tools you need to get your message across.

Croley: I want to underscore the unique role we have here at Motorola Solutions as the innovation and design team. We are a cross-functional team entrusted with putting new advanced-concept visions in place and carrying them through into production. The real magic happens when we listen to the customers—really truly listen. And always we keep our minds broad enough to envision what lies beyond the horizon.

Matthew Scholz

Researcher

3M

Matthew T. Scholz *graduated from the University of Michigan with a bachelor's degree in chemical engineering. He has a master's degree in chemical engineering from the University of Minnesota. He has been with 3M Health Care for 30 years, working in the Orthopedic Products Division and in Infection Prevention Research and Development. Scholz has been integrally involved with the introduction of numerous 3M Healthcare products, including:*

- *Scotchcast Plus casting tape*
- *Scotchcast splints and cast padding*
- *3M Ioban EZ Incise Drape*
- *Avagard D instant hand antiseptic*
- *3M Avagard Surgical and Healthcare Personnel Hand Antiseptic*
- *3M Aseptex fluid-resistant molded surgical mask*
- *antifog/antireflective face shield masks*
- *3M Skin and Nasal Antiseptic*
- *Scotch Blue painter's tape with Edge-Lock*

Scholz is a co-inventor on more than 115 issued US patents. He is currently a corporate scientist in the Infection Prevention Division Laboratory, working to develop products that reduce the risk of acquiring an infection in a health care facility. Total cumulative sales of these products are $2.5 billion.

Scholz has won numerous awards at 3M, including the prestigious Carlton Award—3M's highest recognition of scientists who have made extraordinary contributions. He is the proud father of three children (twins Sam and David, age 16, and Katie, 24). When he is not coaching or attending their school and sporting events, he is likely exercising somewhere. He is an exercise fanatic and an avid runner, having completed the Twin Cities Marathon three times.

Brett Stern: Tell me about your background. Where were you born, and what was your education and field of study?

Matthew Scholz: I was born in Elmhurst, Illinois, just outside of Chicago. I lived in different suburbs of Chicago until I was thirteen, and then I moved to a suburb of Detroit. I went to the University of Michigan, where I got a bachelor of science degree in chemical engineering. Then I took a job with 3M. Two years after working at 3M, I decided to go back and get my master's degree in chemical engineering from the University of Minnesota. I went to school part-time while working full-time.

Stern: When you were a kid, were you inventing things?

Scholz: Yes, actually I did. I was a gearhead. I souped-up cars and had a lawn-mowing business. I always modified the lawn mowers so that I could get more done faster, and pick up the grass better. I even modified the blades of the lawn mowers. I put wings on them before they ever had wings on them to get more grass in the bags and that sort of thing.

Stern: The engineering field was something that you knew you wanted to go into.

Scholz: I always liked math and science, and I really liked mechanical stuff, so this was it for me. I have been happy ever since.

Stern: I looked at your bio and you have a number of patents. Could you explain some of your favorite products that you have invented?

Scholz: It's hard to decide which are my favorites, but certainly the casting tape is probably the most well known. The state of the art before 3M got involved was the plaster of paris cast, which is heavy. And you can't take an X-ray through it. It breaks down in water. Even without water, it breaks down. At 3M, Don Garwood, a guy before me, invented a water-curable synthetic casting tape, but it was very sticky and difficult to apply. In talking to customers, we found they really wanted something that applied like plaster of paris, but had strength and resilience, and allowed you to take an X-ray through it like synthetics. They wanted those properties to combine.

It is a little bit of an oxymoron because what we really needed to do was come up with a slippery adhesive. People don't think of adhesive as being slippery. We called it Scotchcast Plus, and that pretty much dominates the market today. If it is not our product, it is a product that is just like it.

Stern: You worked for 3M, which is a company known for being innovative and developing new ideas. You started years ago—were you given a career path in your first position?

Scholz: I was really quite fortunate. I worked for a very innovative boss. The wonderful thing about 3M—and I really can't explain it until you come here and experience it for yourself—but I can honestly say in thirty years at 3M I have never, ever been turned down for help. And believe me, I have asked hundreds, if not thousands, of people for help.

Stern: Help when there was a mechanical or a technical problem type of thing?

Scholz: Chemical, mechanical, or if I needed time on a process line to prove out a point. The other day, I got a call from a colleague who said, "I heard you were working on this. I think the process that I am working on might help you guys. Are you interested?" I said, "You are darn tootin' I am interested. I would like to give it a try." So the neat thing about 3M is that we have experts in just about every field that you can possibly imagine. We have got all these people that are willing to help you. But, you need to take advantage of it. There are a lot of people at 3M that take advantage of it and there are others that don't.

Stern: Can you sort of talk about the philosophy or the mission statement of how this happens? What makes 3M a continuously innovative company?

Scholz: I come here to work every day and I am passionate about making lives better for patients. You are only on the planet for a short period of time to make a difference. There is not a better place in the world I can think of making a difference than here, because I can get so much done in so little time. If you take a look at that list of patents I have, you will find very few of them that have me listed as the only inventor. There are always co-inventors.

Stern: So you work in collaboration?

Scholz: Absolutely.

Stern: If you have an idea or someone else has an idea, do you go around looking for someone that can work on it, or does management team you up?

Scholz: I personally never go around looking for somebody that can work on it *for* me. I look around for people that can work on it *with* me. I am not going to ask somebody to go invent something for me, but I will ask him or her to invent something with me.

Stern: When you are doing the collaboration, you are working with someone who has a different skill set. So how do you define who is going to be the leader or divide tasks? What is that process?

Scholz: Until it becomes a formal program, it is not really ever discussed. It just kind of happens spontaneously. As an example, there are a couple of programs in 3M where you can get corporate money to pursue your idea. So, if I have an idea that is outside of my division and doesn't really fit in with anybody else's division, I can get corporate dollars to pursue it. There are two programs we set up to do that. One is called Discover, and one is called Genesis. It is just a loose group of people that get together saying we are interested in this idea. You go to the meetings, and people volunteer to get stuff done. People work at that because it is fun. It is not a situation where you have to define accountability measures.

Stern: When you are starting a new project, what is the process that you go through? How do you get from the concept to the finished idea? Are you sketching on a napkin or are you singing in the shower?

Scholz: All of the above, and then some. I keep a pencil and paper by my bed, because a lot of times, I have trouble sleeping at night, so I write down ideas. Initially, if I can do it myself, I will try it in my lab. If I can't, I will find somebody that has the equipment that I might need to prototype. A number of my co-inventors are machinists, so we have machine shops. I will go to the machine shop and ask for them, but I never just go down to the machine shop with a blueprint. First of all, I am not very good at drawing, but secondly, those guys are artists themselves. If I explain to them what I am trying to do, they will make it better.

Stern: It sounds like you're not really doing a lot of modeling but getting hands-on with it right away.

Scholz: Yes, I try to get hands-on with it right away. If it is a complicated product, we may do some modeling. I do a lot of chemistry too. Usually I just jump into it.

Stern: You just listed a few skills that you are not good at. Could you tell me some skills that you are good at?

Scholz: I think having the engineering background and the basics of knowing chemically what is going on. That is the strength that I bring to 3M. There are a lot of people that are great engineers and a lot of people that are great chemists, and I am able to bridge the two.

Stern: Do you have a daily routine that you go through to solve a problem?

Scholz: That depends on the day. I always work on multiple projects at a time, so each day can have a focus. I go from one problem to the next, or one challenge to the next I should say, but it is not something I normally map out.

Stern: How many projects to you work on at a time?

Scholz: At least ten.

Stern: Why so many?

Scholz: Because I have broad interests. I am a corporate scientist. I work on stuff for other divisions as well. People call me up and ask for help, and I am interested in trying to help them with their problems. I am a very driven person, so when I am at my kids' swim meets, I am not just watching the swim meet. I am usually reading patents, or literature, or something when my kids aren't swimming, but I never ever miss one of their races.

Stern: You have been at 3M for a period of time. I am sure you have had the opportunity to be in a management position. But, it sounds like you are still behind the desk mixing things together?

Scholz: Yep. Management would be a fate worse than death for me.

Stern: Why is that?

Scholz: Because I don't enjoy budgets. I don't enjoy the personnel issues. I do have some people reporting to me, but I am more of a coach and a mentor than I am a boss. Thank God for the people around me that like to do all the documentation, because it is not what I love to do. I love to invent stuff, and that is my job. There are people that like to do the other stuff.

Stern: There is no directive that you have to go into management?

Scholz: No. Absolutely not. 3M has a dual ladder, and I do a talk on that every year—along with a friend of mine that is in management—to new employees to let them know that there is a dual ladder.

Stern: You talked briefly about mentoring. Do you still have mentors?

Scholz: Yes. I still keep in contact with several people that I would say are my mentors. The guy that has mostly been my mentor throughout most of my career is retired now, but I still keep in contact with him. He helped me with a lot of chemical problems throughout my career.

Stern: When people are looking for mentors, what would you suggest they look for or how to approach a person?

Scholz: People ask me that question all the time. First of all, I think they need to be a couple of levels above you. Try to get someone outside your immediate area and that complements your skill set. If it is not working, find a new one, because there are too many people that try to make these mentor/mentee relationships work. Sometimes they work, and sometimes they don't.

Stern: When you are starting a new project, do you have any particular process that you use to define a problem or use as a starting point?

Scholz: I always try to boil it down to the basics, and I always try to think about how other people have solved similar problems. When I was working on the casting tape, I asked myself, "How do other people get rid of stickiness?"

You know, oil comes to mind. I think out of the box. I like to think about how Mother Nature would solve this problem. How would an insect solve this problem? How would a child solve this problem?

Stern: So, your inspirations come from everyday situations?

Scholz: A lot of times, yes, but I also get a lot of inspiration from reading literature. I spend a ton of my time reading articles.

Stern: It is interesting you say that, because a lot of people think an invention is something totally new. But in reality, it is all based on prior art from somewhere.

Scholz: Well, these days there is just about prior art for everything. You are absolutely right.

Stern: How do you balance "doing something new" with "just a little basic improvement"?

Scholz: That is easy. I ask my customer. If my customer does not think it is a "wow" idea, I don't even work on it.

Stern: Who is your customer?

Scholz: It depends on the product. It could be a doctor, nurse, or infection preventionist.

Stern: Are you the guy who invented the blue tape?

Scholz: No. I didn't invent the blue tape, but I invented the blue painter's tape with Edge-Lock. Again, it was a collaborative effort with several other people as well.

Stern: Going back to looking for the customer, do you go into the operating room or into the doctor's office and talk directly to the doctors, nurses, or the patients?

Scholz: Absolutely.

Stern: What type of feedback do they try to give you? Do they give you solutions? Or do they give you the problems?

Scholz: Sometimes they give you the problems, and sometimes you have to observe the problems. I am really looking for problems that they aren't even aware that they have. So I am looking for that one person, out of a hundred customers, who is so frustrated with what he is doing that he can express it, or she can express what is going on. It might be cutting a product to make it work or something. There is a whole science behind that called "lead user," which was developed by a guy at Harvard. I try and use it in every project I work on.

Stern: Do you think that if people don't understand something, they won't want to accept it?

Scholz: I think people don't know how to use their imaginations. As an example, until you see an iPad, you can't imagine it. You just deal with things the way they are. You work around a problem, because you can't even imagine that it can be any better.

The other thing that I think I bring to 3M is that I have a good memory for technology. I can go to several poster sessions[1] a year and see hundreds of different new technologies, brand-new-to-the-world technologies. When I go out to my customers to see the problems they are having, I can marry those two: technology and their problems.

Stern: What skill sets do you think an inventor needs to be successful?

Scholz: They need to be able to align the technology to solve the problem. They need to be aware of what I call "the toolbox." They need to have a big toolbox. If all you've got is a screwdriver, everything looks like a screw. You've got to have a big toolbox to be able to really impact a marketplace.

Stern: When you are doing these projects, do you ever get to the point where there is just failure? How do you get over that?

Scholz: Absolutely. I think every project that you commercialize goes through a lull. That is why I tell people, "You need to invent methodically." For every project I am working on, I have a backup plan. If Plan A fails, I have to have a backup plan. Because nine out of ten times, Plan A fails.

Stern: Is it just human nature to pick Plan A, and then just force it to work?

Scholz: Surely.

Stern: What is your mental process to evaluate problems or failures?

Scholz: I force myself to do it because I have been through it so many times that Plan B, C, or D actually is better than Plan A. That is why I work on at least ten things at once. It is probably more than that. It is probably double that because I have always got other people in the company or outside the company—I work with a lot of people that are outside the company too—working on things.

Stern: Have you had a big failure that you look back at and say, I really messed up?

---

[1] 3M's Tech Forum is an internal organization that enables technical interaction at the grassroots level, no matter where technical employees are located in the world or in what capacity. The organization sponsors an annual Spring Symposium and the Tech Forum Annual Event. Both events bring global scientists together into poster sessions, where proprietary research is shared. This interaction spurs innovation and fosters collaboration in technology development.

Scholz: I still get jabbed about the casting tape. The first time I ever brought it over to marketing to show it to them, it was incredibly sticky. They are still joking with me about how sticky it was the first time I showed it to them. It was supposed to be a display of this great slippery casting tape and show how easily it would apply—and it was flypaper.

Stern: What did you do to fix the problem?

Scholz: On Friday, it was very slippery, and on Monday, it was sticky. I had to go back to the drawing board to figure out what was going on. I got help from our corporate analytical people and figured out what was happening. That failure actually made a better product. I was actually able to make it much more practical than I would have otherwise.

Stern: Let's say you come up with a solution that is not patentable. Do you have to keep going until you get a patent on something?

Scholz: No, not necessarily. If the customer thinks it is a "wow," we will market it with or without a patent.

Stern: Do you have to go with any particular size of market? Let's say a doctor comes to you and says, "we could really use this during this procedure," but they may do only one hundred procedures a year. Do you have a point where 3M says an opportunity has to be a certain size financially?

Scholz: Absolutely. My first customer is my business department. If I can't sell them on it, there is no way I will give it to the customer. 3M always tells people to think outside the box. When I get talking internally to technical people, I tell them to think "outside the box inside the box." You need to know what box you are working in, because if you don't have the salesperson to get it to the marketplace, it will never happen.

Stern: Do you get involved in the commercialization side?

Scholz: Yes, absolutely. I go to the trade shows. I talk to the customers. I see what they are doing. The most fun part of being an inventor is to see the customer use your product. That is the best part of it, to see a customer use it and see their eyes light up. That is why I do this.

Stern: So, it is not necessarily financial rewards?

Scholz: There is no financial reward at 3M for a patent. My financial reward is the same with or without the patent. No additional compensation.

Stern: Are there any professional heroes that you have in your field?

Scholz: Now that is a good question. There is a retired 3M man named George Tiers. I would say he is definitely one of my heroes. He epitomizes 3M's culture of sharing, and I will never forget one of the first conversations I had with him. We were talking about going to one of those poster sessions I talked about. Usually there are between one hundred and two hundred people, and he says, "Matt, I go to the poster sessions not to see what I can get from each poster to use on my projects, but to see how much I can help each one of those people." That really epitomizes George, and he really had an impact on the way I try to do my career.

Stern: Are there any inventions that you wish you thought of?

Scholz: Sure. I wish I invented the lightbulb. In my heart, I wish 3M did more implantable stuff, but we are not an implantable company. I think 3M has a lot of biomaterials expertise that we could bring to those fields, but so far, we are not that company.

Stern: Is there a particular part of your field that you really like going into or that you want to explore besides the implants?

Scholz: I am passionate about trying to make sure that people don't get an infection in the hospital. If you see what these people go through, it is absolutely devastating. Imagine getting a hip implant and then a month or two later finding out that you have an infection that is so bad, they have to take you back into surgery to take that implant out, put you on IV antibiotics for two months, and then try to put the implant back in. So, you have gone through three major surgeries. You may not survive.

Stern: Is that a field that you work on continuously?

Scholz: I am trying to prevent patients from getting infections.

Stern: Do you have any method that you use to motivate yourself? How about motivating your team?

Scholz: First of all, I am not by myself. I work closely with a team of people. I have technician help all the time. So, to motivate myself I have always got an active to-do list, stuff I want to get done. I always try and prioritize what I can do here versus what I can do at home or on a weekend. I want to spend my valuable time here at the laboratory. I generally don't read articles here. I read them on my own time. I am trying to get done with what has got to get done here. From the team's standpoint, I try and take a look at the projects and their scientific approaches, and figure out what I think is most likely going to work and prioritize that first. But again, we always have backups, so we are always looking at backup plans.

Stern: Are some of your projects that you are working on five, ten, twenty years down the road?

Scholz: I try not to work on things that are that far down the road. I try to work on projects that I can get done in less than five years.

Stern: You are planning on working for five more years at least?

Scholz: Probably at least ten.

Stern: Do you have a plan to retire?

Scholz: Not any time soon.

Stern: What do you plan to do when you retire?

Scholz: Probably do volunteer work. Work at Habitat for Humanity or something like that. I do enjoy talking to classes and I do a lot of what we call "visiting wizards" in the classroom and promoting science in the community. I love kids. I have three of my own.

Stern: What do you ask kids in the classroom?

Scholz: I ask them to be really passionate about trying to get it done and make sure they get it done. Be very clear on what you want to accomplish. You have got to find a way to get it to work.

Stern: What about people that are not necessarily good at math or science but have a good idea. How do they go about doing that?

Scholz: The same way you do if you are good. You've got to collaborate with people. You have got to get people to help you. You are never going to bring a product to market by yourself, so you've got to collaborate with people. If you read the Steve Jobs biography, you see he was not particularly good at science or math, but he was a good "idea" guy who had his team together to make it happen.

Stern: Outside of the products that you are developing, do you have a favorite product that you like having around, either technological or nontechnological?

Scholz: I just couldn't work without my computer, I will tell you that. I am near and dear to my lawn mower because I had a lawn-mowing business.

Stern: Do you still mow your grass?

Scholz: I still mow my grass, yes. I still fix up lawn mowers and give them away to people that need them.

Stern: Well, do you have inventions in lawn mowers that you think you want to bring out?

Scholz: No, I don't think so. I modify them and I fix them up, but it is not my passion. I want to spend my time trying to help patients, so I usually do it that way.

Stern: Obviously, you are going to the marketplace looking for problems. What about if doctors or consumers create some possible solutions. Does 3M open up their doors to let those ideas come into your sector?

Scholz: I get several every month coming in, so yes.

Stern: What is the process that you go through to evaluate them, and do they ever actually come to fruition?

Scholz: They rarely come to fruition. People are always overestimating the business potential. There is a formal process. 3M has to protect itself, first of all, and I can understand why, so we always have to sign a confidentiality agreement first and make sure they realize that 3M is probably working in this area and already considered several iterations. We prefer that the inventor have a patent application on file, but that is a pretty big hurdle. A good application might cost you $10,000. That is a pretty big financial hurdle.

Stern: Any words of wisdom that you want to give to inventors out there, or young people?

Scholz: You have got to have the drive. If you don't have the drive, you are never going to get it done. I think that is more important than your science and math knowledge. It is far more important. And you have got to network with people. Those two things I think are the biggest. If I were going to hire a dream team to work on something, I would hire first the people that are really passionate about getting something done and then people who collaborate well.

Stern: Why is the networking so important?

Scholz: Because everybody brings a different outlook and has a different skill set to get it done. You can get it done way faster with people from diverse backgrounds. I work with people all over the world at 3M. There is 3M China's lab, Brazil, Germany, Japan. You know we have researchers everywhere. Researchers look at problems differently. It doesn't really matter what country you are from.

Stern: So the innovation process, it does change somewhat from people all over the world?

Scholz: I don't think it changes by where you are in the world as much as what your background is and what your experiences have been, in my opinion. We have had people here from all those countries that I have mentioned, and I don't see any trait that I would say is different about the person from Asia versus America or whatever. It is just that the people that have the drive are the people that get these projects done.

Stern: When you are at a dinner party, what do you tell people that you do?

Scholz: I tell them I work on products for health care.

Stern: Are there any products that you want to talk about that you are working on now?

Scholz: I really can't talk too much about anything I have that is not commercialized yet, but I am working on a number of products to help reduce bacterial infections that don't involve antibiotics.

Stern: What do you do for fun or distraction?

Scholz: I am a fitness nut. I love to run and swim. I love to go out with my kids. Those two things probably take up ninety-nine percent of my time. I work out every day. I love to work out. I am a fanatic.

Daria Mochly-Rosen

Professor, Chemical and Systems Biology
Stanford University, School of Medicine

Dr. Daria Mochly-Rosen is a professor in the Department of Chemical and Systems Biology, the Senior Associate Dean for Research, the founder and co-director of SPARK At Stanford program and the George D. Smith Professor of Translational Medicine at the Stanford University School of Medicine. She leads a multidisciplinary research lab that includes chemists, biochemists, biologists, and physician scientists. She has used her basic research discoveries to develop a number of drugs for human diseases. She has been issued 16 patents to date.

Dr. Mochly-Rosen applied her basic research training in protein chemistry to understanding signal transduction in normal and disease states. Recognizing the therapeutic potential of some tools that she has developed in basic research, she spearheaded the development of a novel category of therapeutics for human diseases. These "first-in-class" compounds were the basis to her cofounding KAI Pharmaceuticals in 2003, which was acquired by Amgen in 2012.

Recently, her lab identified new drugs that regulate a key enzyme of cell survival under oxidative stress. These drugs (called Aldas—short for aldehyde dehydrogeanse activators) show great promise in animal models of acute myocardial infarction, diabetic complications, and neurodegenerative diseases. The therapeutic potential of Aldas led to founding in 2011of ALDEA Pharmaceuticals.

Her experience in translating basic research findings to human studies and drug companies led her to initiate SPARK At Stanford, a university-wide program she co-directs. SPARK helps inventors of biopharmaceuticals and diagnostics bring their invention to patient care. This unique program provides faculty education in drug development and supports the school's mission in education and in translational research.

Brett Stern: Can you give me your background? Where you were born, your education, and your field of study.

Daria Mochly-Rosen: I was born in the north of Israel, close to the Lebanese border. I am the only scientist in my family, at least among my siblings and my parents. I started my science interest in high school and actually worked then on a project that eventually offered diagnostic testing for everybody born in Israel.

I did my PhD in the Department of Chemical Immunology at the Weizmann Institute after doing my undergraduate at Tel Aviv University. I came to the United States in 1983 to do my postdoctoral training in the Department of Biochemistry at the University of California, Berkeley. I started working on enzyme kinetics and the like, and continued the project at the University of California, San Francisco. But only when I came to Stanford did my research get to a point where I started to think that the tools that we generated for answering biological and biochemical questions might have therapeutic utility.

Stern: Could you give a brief explanation of what your research covers?

Mochly-Rosen: Our research focused on a family of enzymes called protein kinase C [PKC], which are little molecular machines whose functions include regulation of gene expression, cell death and other responses to insults, and regulation of hormone responses and ion channels, to name a few. The enzymes in this family are highly homologous—they all look very similar to each other—and are called PKC isozymes.

What puzzled me is why we evolved to have a family of enzymes that are so similar to each other. It was natural to assume that they can't be doing the same job—that each isozyme must have unique functions. To sort it out, isozyme-selective inhibitors had to be generated. The effort of quite a few drug companies focused on inhibiting the machine or active part of each isozyme. That turned out to be difficult, because the kind of inhibitor drugs that they made affect not just one member of the PKC family but all its members—and, in fact, many other families with similar functions. So with these drugs, it was not possible to sort out the distinct functions of each of those members of the PKC family.

We started to think that perhaps the way we could sort out the functions of the various PKC isozymes was to look at where they are in the cell. We reasoned that they are each compartmentalized in a different way. We indeed discovered that the PKC isozymes translocate from one cell compartment to another when activated by the appropriate signal, with each isozyme translocating to a unique subcellular site. Then we had the idea that location determines the functions of a given isozyme. If each isozyme has a different "zip code," and we could block access to that address, we would deactivate the function of that isozyme without affecting the function of the other isozymes.

As this idea was very much against the dogma at the time, we had to really go the extra mile in our work to prove that our theory was correct. We first proposed and then showed that there are corresponding isozyme-specific anchoring proteins, called receptors for activated C-kinase [RACKs], that provide the address for each of the isozymes. We then started to develop tools to inhibit the ability of the isozymes to bind with their corresponding isozyme-specific RACK. These tools were generated to prove our hypothesis, but soon after it became clear to us that they can be used also as drugs, to inhibit isozymes that contribute to specific pathological conditions.

Stern: You have worked in an academic setting for your entire career. Is there a reason for that, as compared to a corporate setting?

Mochly-Rosen: Yes, I'm a curious person, and going through academia and doing research in an academic setting allows me to follow my curiosity. I can decide on the topic that I want to study, set a hypothesis, and explore it, provided that I get funding for it. And, if I find something else that is interesting along the way, I can change my mind, switch gears and move to work on another problem. This freedom in the academic setting to follow my curiosity is extremely liberating.

At the same time, I'm really a "people person." I like to teach and I like to learn. So these are the reasons that academia suits my work and personality.

Besides, I had very little knowledge of what's done in industry. I had not been exposed to it before, and so I just thought it's not my world. Only after I spent a year in the company I founded did I realize that I had a lot of misconceptions about working in an industrial setting—misconceptions shared, I might add, by a lot of academicians. Still, I had no doubt that I would come back to academia, which is really my world.

Stern: Going back, how did you start this research project? What was your inspiration to see this problem?

Mochly-Rosen: It's funny that you say, "see" because I'm a visual person and I actually did see it. I looked in the microscope and I saw something, and it was clear to me that what I saw could be explained by this idea that the location of an isozyme determines its function.

Stern: So you just looked through the microscope and had that Aha! moment?

Mochly-Rosen: Yes. And I was surprised, because in retrospect, quite a few other people were looking in the microscope at that time—but somehow they didn't put it together that way.

Stern: What were you actually looking at, at the moment?

Mochly-Rosen: At the PKC isozymes in the cells of a beating heart.

Stern: So after looking through the microscope, did you just sit down and write something out?

Mochly-Rosen: I am a visual person, so I drew something to represent my idea that the isozymes anchor themselves at these various places by binding with those isozyme-specific proteins that I called RACKs. I thought, "I need to present and explain my idea to others, but I'm worried they won't get it." I'm a foreigner—as you can tell. So I drew a cartoon of hats of various forms to represent the isozymes, and then racks of various forms to represent the protein anchors, each of which fitting for only one type of a hat. That drawing in my lab notebook was copied by a co-worker and is now framed and hanging in my office. It's very faint now because it's from 1988.

Stern: So this is a problem that you've been working on continuously since then?

Mochly-Rosen: Yes. Actually, just last week I was awarded another NIH [National Institutes of Health] grant to continue with this project. It continues to evolve, but the initial idea led to a rational design and production of drugs that we use to investigate the RACKs. In the beginning, it was very difficult to convince people about the importance of this finding so we decided to apply these drugs to living heart cells that beat in the dish. We then showed that by changing the location of specific PKC isozymes, we changed the beating rate of the heart cells. So we showed that our hypothesis was correct—and that it mattered.

Here's a prime example of why the word "we" is appropriate to my work. After we completed the above experiment, I went to the American Heart Association and gave a talk on how the drugs we made can regulate the beating rate of heart cells in the dish, and the crowd was extremely bored and uninterested. Then the chief of cardiology at UCSF, Joel Karliner, said, "You know, this is really nice, but for cardiologists, regulation of heart rate is not important. You need to work on something that we care about, like heart attacks."

I said, "I don't know how to work on heart attack. I don't have any expertise."

He said, "Okay, I'll send you a cardiology fellow."

So Mary Gray came to my lab, and we started to work on heart attacks in the dish. When she left, I continued to have MDs in the group. So my initial idea was a solitary event that was worth nothing in itself. Many people have worked on it for many years, developing it and applying it to interesting problems. The trick is to get out of the comfort zone, talk with others seek advice and work in mixed teams.

Stern: You said that you created a company. What was the reasoning behind that?

Mochly-Rosen: We realized that we had a product with an important potential utility—it can help patients who have heart attack. Stanford has a very good Office of Technology Licensing [OTL], and they shopped the idea. But nobody was interested.

Stern: They shopped the idea to the pharmaceutical industry?

Mochly-Rosen: Yes. The Stanford's OTL reaches to companies that are in the field, yet no one showed interest. This was in 2000. I started to send the reagents to any academician who asked for them. A direct competitor, for example, said that he didn't believe the data that we presented in a meeting, so I sent him the reagents. They got the same data, which made us feel more comfortable.

By the way, let me divert for a second and describe something that has happened to me again and again in my research life that is relevant to entrepreneurship. While we were working in the lab, I was one hundred percent convinced that the idea was correct. I had great confidence in it. But once I would present the experiments or put the paper for publication, I would start to worry that maybe the idea is incorrect.

That's why it is always very important for me to give the reagents to anybody who asked. And to describe in detail how we did everything and invite them to our lab to do their experiments. This was a way to make sure that our data were correct. I didn't fall in love with the idea to such an extent that I lost the ability to be critical of it.

That is important when it comes to your question about starting the company. So the OTL was doing its job, but was not successful. The papers were out. We were giving heart attacks to pigs resulting in human-equivalent of four hours of complete loss of blood flow to large areas of the heart. And giving the pigs our drug after the heart attack reduced the injury to the heart by seventy percent. You give a drug after a heart attack and it conferred a big benefit.

You'd think that somebody would think that is important drug, right? I thought the Office of Technology Licensing was not doing a good job there, so I should do it. I talked to lots of people and I got to know quite a few people in Big Pharma and start-ups. I started to realize that OTL was doing a terrific job; the problem was that our project was missing important elements for pharma to get interested.

It was very much against their dogma in every aspect. The drug was not the kind of drug that they were making. It was a peptide—a tiny fragment of a protein and those were not considered to be good drugs. The drugs also worked by getting inside cells, so we found a way to deliver the peptides inside cells, which pharma was not used to, and so on.

Stern: What was the process that you went through to start a company? Here you are a scientist, but now you're going to be a businessperson.

Mochly-Rosen: There are more twists to the plot, because you need all these ingredients in order for things to happen. The last ingredient was my graduate student, Leon Chen. He kept saying, "Let's start a company."

I said, "Absolutely not. Not interested. I'm not an entrepreneur. I don't know how to do it. I'm an academician. I don't care about this kind of stuff. Somebody who knows how to do it should do it. I'm not doing it. Eventually something will pan out with Big Pharma."

He said, "And if not?"

I replied, "If not, by the time you graduate, I will start a company."

That was a stupid thing to say, because Leon graduated in record time—one of the shortest PhDs in our department's history. So I kept my word; he graduated in December, and so in January I bought my first notebook and I started to talk to people about starting a company.

Stern: Not having any business background, what was your methodology to start a company?

Mochly-Rosen: I talked to hundreds of people. I had four pictures at the end of my notebook: "This is your brain on drug" kind of pictures. It showed a brain after stroke without the drug and a brain after a stroke with the drug. It also showed a heart after a heart attack without the drug—and one treated with the drug. These four pictures told the whole story—they were unequivocal and a great "elevator pitch."

Stern: Were you asking questions related to business, or were you asking questions related to the science?

Mochly-Rosen: I asked questions related to anything that they had to offer. I always had the notebook with me and I flashed the four pictures to anybody who talked to me about anything. Then I'd ask them if they had any idea what I needed to do next. I would talk to clinicians, researchers in high tech, other faculty, the neighbor when I threw out the garbage—from whom I got one of the most important introductions to a corporate lawyer—and people on the plane. It was a long process. And I was patient. And, again, I didn't have any ego about that. It was really about getting the stuff done.

One of the things that happens when you start to talk to people—especially when you start to talk to venture capitalists and Big Pharma—is that you have to realize that people will tell you no. I didn't get many "nos" in my career until then. So if you go into it with your ego exposed, you'll get crushed.

Stern: Before you said, "You do what you know." What have you learned now as a businessperson, as compared to being a scientist? Do they conflict with each other?

Mochly-Rosen: They don't necessarily conflict. It's just that they're different universes. I think that the biggest problem that we have is an issue of what words we use when people from these two universes talk to each other. If you talk to me about a field that I know nothing about, and you use terms that I have never heard, I know to ask you for its meaning. But if you use a word that I understand, but my use of the term is different from yours, we have a problem understanding each other.

Stern: So you try to get a common dialogue?

Mochly-Rosen: Yes. It took some time. I needed to ask all the time for clarification. And when I spoke, I always rephrased what I heard in different ways to make sure we were talking about the same thing.

Stern: In the businesses that you have started, have you brought a product to market?

Mochly-Rosen: Not yet. One company—KAI Pharmaceuticals—was recently acquired and it remains to be determined whether any of the research from my lab will be translated into a drug.

Stern: So was selling this company in a sense a way to license the ideas? Was the business sort of a format to commercialize your ideas, but not necessarily bring the product to market?

Mochly-Rosen: I think everybody is hoping to bring product to market, but there is no way—or at least it's extremely rare—that a start-up biotech company ends up bringing a product to market itself. Of course, we have the stories of Genentech and the like, but they are distinguished by their rarity. Usually a start-up brings the experimental product up to Phase 2 trials and then it either licenses it or the company is acquired by Big Pharma to do the large clinical trials. That's exactly what happened to KAI.

Stern: You run a program called SPARK At Stanford. Could you tell me what that is and explain it?

Mochly-Rosen: SPARK is a program designed to foster partnerships between scientist entrepreneurs and industry experts and facilitate the transition of research discoveries from Stanford laboratories to patient care. The inspiration for the program was what I learned during the year I spent raising funding for my first company, especially about why we were being turned down—about what was missing from our work that VCs and pharma people were looking for. I have learned even more about the importance of what industrial experts do and how they do it—and how we in academia are absolutely clueless about it.

I returned to Stanford exactly a year after I left to start up the company. When I got back, I thought that I had learned a great deal that may help others.

When I started the company, I wanted to get it right. I sought advice from other entrepreneurs at Stanford, but the main advice they had was to warn me to watch out: the world outside is harsh, everybody is taking advantage of you, and also—it's not intellectually interesting. When I spent that year "out there" in the industry, I verified that it is indeed a harsh world, but I also discovered that it actually is intellectually interesting, challenging and very satisfying. I thought that this was worth telling others and giving an opportunity for other inventions to mature enough so that they will move into the hands of pharma.

Also, many of our students and fellows are going to end up in that world, yet we were doing very little in the way of teaching them about it. Teaching more about it is part of our academic mission.

With the encouragement of the Dean of the Medical School, Phil Pizzo, I started the SPARK At Stanford program.

I spent some time trying to get money from industry to pay for it. During that year in my company, I had made a lot of friends, and I thought that some of them would be willing to put money in SPARK. They agreed that the program is really needed, but they also said that they absolutely cannot fund it. Instead they offered to help me in other ways. So I created a very long list of people who promised that they would help, and then I went back to them and said, "Okay, can you provide free advice to our inventors?" For six years now, many of them come every Wednesday night—our SPARK meeting night. They signed a confidentiality agreement and give free advice and help our inventors to mature their research and ideas to the stage where somebody else can take it and either license it, or start a company around it, or even just implement it into their clinical practice.

Stern: So it's a situation where a student or a professor could stand up and make a presentation, and get feedback for their work.

Mochly-Rosen: It is a situation for faculty members, who have invented and patented something, but have not succeeded in getting their patents picked up. It's mainly faculty members from the School of Medicine, but we have also faculty members from the schools of Humanities and Sciences and of Engineering, too. We bring these inventors to a competition and guide them how to present their ideas in an industrial setting, which is very different from teaching ideas in an academic setting. And if they are selected for the program, they get not only money to support that effort, but, more importantly, advice that is critical to highlight the value of their invention.

Stern: What are some of the positive examples that have come from this?

Mochly-Rosen: Right now, we have thirty-six active projects. The inventors work using the structure of industry standards. In other words, we discuss the plans and agree on milestones and time line and so on. They present their progress every three months and get real-time advice throughout their work. Since we started, half of the projects either have been licensed or have advanced into clinical trials or both. The other measure of the program's success is education: we have taught over a hundred faculty members, students and fellows the translational process.

Stern: Is the education that your program provides mostly on the business side?

Mochly-Rosen: No, the business side is the topic that we cover the least. We do have a couple of business lectures: one on how to pitch to a venture capitalist and the other on how to evaluate a market. We also have a lecture on corporate law. But the main thing that we teach is how to develop a drug or diagnostic. What do you need for the clinical trial design, the quality assurance issues, and the intellectual property around it? We're teaching how to move from idea to a drug or diagnostic development projects that will appeal to industry.

Stern: What do you think are the skill sets that a scientist or an engineer needs to move a project forward?

Mochly-Rosen: I think that the first thing that they need to know is what they don't know. In fact, we are putting the finishing touches to a SPARK book that basically lists all the things that academics don't generally know about diagnostic and drug discovery and development, who to ask, and what issues to think about. So it's not a how-to book. Each step in the process of diagnostic and drug development is covered in the book with a view to highlighting those features that we in academia are most likely to be surprised by, clueless about, or find counterintuitive. This is also the main thing that we discuss during the sessions. Then we teach things that work and, more importantly, things that don't work. A lot of the learning comes, of course, from failure, but if a project fails in industry, the investigator is unlikely to work on this project again and so the lesson learned are lost. In SPARK, these experts have an opportunity to share this learning.

Another advantage is that when we fail in academia, we still write papers and talk about it, teaching what we learned from this failure, etc. So if we move some of these development efforts into the academic setting, we have a chance to improve the process to benefit the industry, too.

Stern: When you're doing your work, where do you seek inspiration or solutions?

Mochly-Rosen: My inspiration is from the physical world around us. I find that many scientists think about their work in purely conceptual terms. But for me, it's a physical space where three dimensional molecules move in crowded space that makes up the cells. Therefore, very much as our world around us, I conceive this micro- to nano-world as a place with machines and objects that move in space and touch other objects after overcoming physical barriers. The solutions that I see in the macro-world I then extrapolate to the nano-world.

Stern: It sounds like you are trying to find an analogy: something that exists in one scale, and then just change it to a different scale.

Mochly-Rosen: That's right. And scalability is not abstruse. It is a reality. The nano-world is a physical space with the same mechanical, energy and chemical rules as our macro-world. Therefore, if you think about a problem in this scalable sense, you can find a solution.

The other salient characteristic of my approach to problem solving is that I'm confident that every problem has a solution. We are looking to discover a certain truth. It's not a question of opinion. It's a question of discovering how something works. So if you have the confidence that there is one solution, then you work on it with a different kind of energy than if you think, "Oh, this is a mystery and maybe I'll get lucky and figure it out."

People in my lab know my working principle: "It's not a question of opinion. There is an absolute truth and we have to find what it is."

Stern: I assume your goal is eventually to have these products put out there in the marketplace. To what extent does financial reward drive these efforts?

Mochly-Rosen: It has not been my motivator. One of the biggest rewards that I had in my life, and the biggest related to translation research occurred when I pushed the return button on the computer to activate the first clinical site to treat the first patient with our drug. I've never experienced anything equivalent to that feeling.

Stern: If financial reward is not the motivation, what is it?

Mochly-Rosen: The quest to find the truth. To play with a problem in your mind and work on it in the lab for years and ultimately to find a solution: that, by itself, is the motivation. The reward from starting a company came unexpectedly, when I attended the first clinical trial training meeting of the company and realized that hundreds of people joined in to work on our project. I can't tell you how amazing that was. To see so many people joining in, dropping what they were doing to work on our clinical trial. That was a major reward.

But I really love my academic life and that is why I never hesitated about returning to Stanford after my year in the company. I love the teaching and I love the freedom to follow my curiosity and to decide when to start or stop working on a problem that interests me. That's my biggest reward. But I also got a lot back from my experience as an entrepreneur, in terms of the excitement and challenge of developing a drug.

Stern: Along the way, have you had mentors? If so, how do you find them and what's the advantage of having them?

Mochly-Rosen: Thank you for asking this question. We all need mentors and we need mentors of different kinds for different times and different aspects of our life. When I came to Stanford, I was assigned a faculty member as a mentor. But I thought I needed another mentor that is outside my department and tenure track to explain the ropes to me. I didn't grow up in the US, I didn't study here and I was not sure that understand the culture. So I found a professor in medicine and asked him to be my mentor and actually, he was the first person I also talked to that January, when I started the company.

I had scientific mentor and a friend from another university that I continue to consult with to this day. She is not only extremely smart and curious, but she cares about the big picture and a good reminder of what is really important.

I also have a mentor who is an artist. We talk about creativity and art and although it seems as though it has nothing to do with science, it actually has been a tremendous influence in providing me advice with how to move forward when I get stuck.

I also have a family and children, so I have mentors who have advised me on balancing my life. I think that throughout your life you have to find mentors covering different aspects of what you do, people to talk to.

Stern: Do people come to you now as a mentor?

Mochly-Rosen: Yes. I tell them something that one of my first mentors told me: "I'll give you an advice. Then you go out and you either follow the advice or not. But even if you don't follow the advice and if you fail, you can come back and ask for another advice; I will never say I told you so." I say the same thing. Then I say, "This is a way for me to tell you, 'Take it, or leave it.'"

Stern: Is there any advice you would like to offer to an inventor out there in any field?

Mochly-Rosen: I find that each invention has its own life and when you are determining if an invention is useful, then it can't be about you. Leave your ego out of it.

Stern: When you're working on projects, do you have one invention at a time, or do you work on several projects?

Mochly-Rosen: No. I am working on multiple projects at the same time. Perhaps too many.

Stern: What's the reasoning behind working on multiple projects?

Mochly-Rosen: That's the wonderful thing in academia. You suddenly realize, "Oh, this is a problem. It's an interesting problem. Let me try to solve this problem." Then you come up with an idea and start testing whether the idea is correct. Pretty soon you find yourself say, "Okay, maybe I have here something and it's worth developing further."

If I were in a company, I would be assigned or take on a particular project—a particular problem. If I found something else that was interesting to me, that I was curious about, I'd have to shelve it. I'm a very curious person and academia allows me to continue to be curious.

Stern: Any thoughts on being a female or foreigner in your line of work?

Mochly-Rosen: I told you how critical multi-modal communication is when you are working in a team with people with different skills. So it's an advantage to be a foreigner, because we are used to explaining ourselves in more than one way because of our accent, because the cultural references that we make are foreign, and so on. The other aspect of being a foreigner that is an advantage is that you hear a lot of comments from people that are culturally nuanced and, being a foreigner, you oftentimes are not realizing that they are meant in a negative and discriminatory. Not being aware of that is fantastic, because then you are not burdened by it. It took me a long time to understand that some of the things that were said to me reflected the speakers' biases. Not understanding what they were saying was a huge advantage. I've been here now for almost thirty years. The most common thing that happens to me when I introduce myself to somebody is that they ask, "Where are you from?" If I say, "I'm from California," they say, "No, where are you from?"

Stern: When you're at a dinner party, what do you tell people that you do?

Mochly-Rosen: It depends, because I think that a lot of people find what I do a barrier. So it depends on the company. Oftentimes I don't say it right away, I try to avoid the topic. If it's people that I know very well, of course they already know. My husband is not in the sciences. He's a writer, so we have friends with different interests. Usually I would say to them that I'm teaching in the university.

Stern: What do you do for fun?

Mochly-Rosen: Science. And I love art. I don't do any. I am passive.

Stern: So you look at art. You like going to museums or galleries.

Mochly-Rosen: Yes, yes. And I love to read. I read a lot. I read probably too much. Maybe I should read more science.

Stern: Do you plan to retire and what will you do if you retire?

Mochly-Rosen: You are the second person who asked me that question this week. I'm starting to worry.

Stern: It's a generic question.

Mochly-Rosen: Yes, I understand. No, I do not plan to retire, at least not soon. I really love my lab and working on SPARK projects, which has provided me with a new horizon. I always set for myself a particular Everest and then I'm climbing towards it. There is always the problem, however, that when you reach the summit, from there you have to go down. So you have to find yourself the next Everest. I think SPARK has been my Everest in the last six years and I can see that I will not reach the summit for a long time. So this is good.

Stern: Any final advice for inventors out there?

Mochly-Rosen: Remain curious.

Martin Keen

Industrial Designer
Keen Design Studio

Martin Keen, *the principal of Keen Design Studio, has created products for many companies, primarily in the footwear industry. In 2003, he patented a hybrid sandal–shoe for competitive sailors and launched the Keen Footwear brand based on the design concept of ultra-practical utility. Keen has developed designs for innovative sports gear, lighting, and functional art. He developed the concept for a line of upright furniture to facilitate creativity and productivity by allowing sitters to quasi-stand at their workstations in studios and offices. In 2012, he launched Focal Upright Furniture.*

Keen was born in Somerset, England. He holds a BA in industrial design from the Ohio State University. He lives in Jamestown, Rhode Island, with his wife and two children. He enjoys sailing competitively.

Brett Stern: Could you give me some of your background? Where you were born, your education, and your field of study?

Martin Keen: I was born in England but only lived there for a year. I moved to Cork, Ireland, with my parents. My father was a pattern maker for the shoe-maker Clarks of England. We came to the States in '71 when I was almost seven years old. My father took a job in Cincinnati, Ohio, working for a company called US Shoe, again as a pattern maker.

I went through high school and then ended up enrolling at Ohio State University, majoring in industrial design, which was something that seemed very natural. I have always been a very hands-on builder of models, creating things with my hands and dreaming up all sorts of schemes and ideas. So it was a perfect fit for me. I thought I originally wanted to be a mechanical engineer.

I actually started studying mechanical engineering, but I realized I really wanted to be a little bit more hands-on, not so focused on the theoretical development of concepts. Fortunately, Ohio State University had a great industrial design program. They were all German instructors there. I've always liked the German process.

Stern: Could you tell me what industrial design is, as compared to engineering?

Keen: I'd say it's a discipline that's half art and half engineering. It combines an understanding of manufacturing and the processes that go into building and mass-producing product starting with product ideation and development.

Stern: You were saying that when you were a kid you invented stuff. What were some of the things you worked on?

Keen: I built a lot of boat models, and balsa and tissue paper airplane models. I was always very meticulous and really enjoyed just using my hands. So it wasn't so much invention, I suppose. At an early age, it was more just really enjoying the process of working with my hands and creating something from nothing. Obviously, erector sets—all sorts of crazy bridges and things like that.

Stern: So the work that you are known for—could you give me some background and explain that?

Keen: Finishing industrial design school, I decided to follow in the industry my father had been in. He wasn't really a footwear designer, but I was interested in sport. I was a runner and a competitive sailor and I went to work for a footwear company in Boston. Saucony was my first job out of school. I really enjoyed the idea of a product that was very utilitarian, in the sense of a tool for running, a tool for playing tennis—an extension of the body.

As a competitive sailor, I was always looking for a competitive advantage. Footwear is not something that comes immediately to mind as a competitive differentiator in sailing, but a moment's reflection shows that superior grip, protection, and fit in footwear can confer a distinct edge to otherwise equal sailors. If you can take your mind off of whacking your toes on the cleats and winches on a boat, then you can concentrate that much more on the task at hand. So that's where the whole idea for Keen footwear started. It was really a sandal that could provide the protection of a shoe, but would still be open and very breathable and let water pass through it.

When I decided to start my own company, I wanted to start from the foundation, which in footwear is the last. The last is the form that shoes are built over—traditionally made of hardwood but nowadays usually made of high-density polyethylene plastic. In established companies, the lasts are just given to designers, who are told: "We need an update of the Shadow 5000 model. We need you to design the new Shadow 6000 running shoe model around this Shadow 5000 last."

The designer has no idea what considerations went into designing the last he inherits. By contrast, I ended up studying biomechanics in depth and casting a lot of feet, so that I could understand the foot form, not just of the New Englanders of Boston, but of people from Africa, South America, from Asia, and Northern Europe. I cast feet from all over the place and compiled a pretty good library of foot form. From that, I developed my own last shape, which became the foundation of the entire Keen footwear line. And that's why, when people put on a Keen shoe, they feel it's a very natural fit.

Stern: At what point did you decide to go off on your own?

Keen: I worked for Saucony for two-and-a-half years. I was getting a little restless, realizing like anyone else starting in industry that if you change and go to another company, you can double your salary. I had an offer from Timberland, just up the road, and another offer from K-Swiss out in LA. I decided to go to LA to check that out for a little while, and I worked for them for another two-and-a-half years. I decided to leave and start my own consulting business, Keen Design Studio. And from 1994 until I launched Keen Footwear in February 2003, I was working as a consultant in the industry.

I had come up with the idea for the product in 1999, while I was consulting as design director for Nautica, which was sort of an in-name-only sailing brand for department stores worn by the yacht club crowd, rather than hardcore sailors. At the same time, I developed this more edgy, useful, protective sandal that definitely didn't fit into the Nautica realm, so I put it to the side and said, "All right. Well, maybe one day I'll develop my own brand."

What prompted me to actually move forward on that was that event on September 11th. I was working for a number of companies at this time, one of them Tommy Hilfiger. I was design director for that brand, but I was not really enjoying it. Mostly, I traveled to Italy to bring back the latest fashions and produce variations on those themes. To me that was not design, it was just copying. And I was really growing rather bored with working with clients like that. So September 11th rolled along and it was Stride Rite that had the license for Tommy Hilfiger at the time, and they let all their consultants and one hundred and twenty employees go. So I was part of the collateral damage. September 11th had a lot of people thinking about what their life was all about.

Stern: So was that a kick in the ass for you then?

Keen: It was. I had a couple young kids at that point. And you have all these wants and dreams, and it's easy to just keep going along in life, working for the man and taking his paychecks, and never really realizing your dream. So to a certain degree, I wanted to get out of the footwear industry. I had been in it for a number of years at that point and I wanted to try something new, but if I was going to get out, I wanted to get out with a piece, as it were, having made my marks. Maybe make a name for myself.

So yeah, I would say September 11th really was a kick in the pants for me. It made me look at my children and think what I wanted for them. What you want for them is obviously what you want for yourself to a degree, which is to be really controlling your own destiny. So yeah, I would say September 11th was a moving point for a lot of people. Whether you are forced to look for something else because you were laid off, or whether it was just a combination of that and your desire to do something, make a mark.

Stern: While you were consulting, would you say your work was collaborative, or were you on your own?

Keen: I would say it was collaborative in the early stage. Obviously, I was working with marketing and sales people to get their input. But their input very often was tainted. I think it's one of the reasons I started my own consulting business was because I just got so tired of the lack of creative time you had working in-house for these companies. You spend maybe twenty percent of your time actually being creative. The rest of the time was sitting in meetings with marketing and sales people, and planning for the future. And that's not what designers are all about. Designers are dreamers and they say, "Okay, give us a basic direction, and we'll create the future for you in 2D or 3D." That's really what I wanted to do. That's why I ended up starting my own consulting business.

Stern: Can you explain your ideation process? What steps you go through to solve a problem?

Keen: My approach is to look at what the current market is and who the current consumer is, and then look more broadly at who the consumer could be in the context of the brand. Let's take as an example one of the brands I worked for—say K-Swiss. I worked for K-Swiss in the early nineties. It was a tennis brand for people who aspired to be tennis players. There were two new avenues that K-Swiss was considering: yacht club aesthetic and country club aesthetic. Yet the core tennis market was really coming along at that time, and K-Swiss didn't really have any technical product, even though they once did. So I started developing a lot of core tennis shoes because I realized what the brand stood for and that it was originally developed as a core tennis product. I really just expanded that heritage.

Stern: So you're starting with looking at the current consumer and then the potential consumer.

Keen: Correct. You look at the heritage of the brand. You scrape everything away and ask what does this brand stand for? So it really comes from brand. All the 2D or 3D concepts are conditioned by the brand's heritage and associations. Starting from that point, you scope cross-markets. They aren't necessarily tennis players, but they want to appear to be by dressing as members of the club, as it were. So that's when I also got the brand into the urban market, which was a huge, booming market. A lot of urban youth aspire to be "members of the club," as it were.

Stern: Do you do a lot of research into the marketplace?

Keen: Yes. First of all, the history of whatever company you're working for. Then, who has adopted their product and who has not. And then how can you expand the marketplace without stretching too much and creating a redheaded stepchild that really doesn't fit in with the heritage of the brand.

Stern: How do you go about thinking of ideas and starting your ideating process? Do you sketch? Do you prototype?

Keen: Yes, I do. I visualize in three dimensions what the product looks like and, being such a hands-on builder of products, I like to very quickly get to the three dimensional form. I spend half an hour on a simple sketch, and then get right to trying to build something, either carving out of foam with a rasp, or building Bondo up onto a last shape. I build a lot of footwear sole and upper models out of Bondo.

I build rough concepts, then refine the sketch, and then go back and refine the 3D form. They are really 3D studies, accurate to the point that I can take them to a manufacturer who laser-scans them. Then the product is built from the scan. So what you buy in the store is the same 3D form that was created by my hand, using Bondo and a rasp. The method I use is unusual in footwear design. A lot of footwear designers became very good at creating pretty pictures and then they trust a model maker or factory to build a model. Footwear realized in this manner often ends up excessively straight-sided. By contrast, I seek to control the 3D form that people pick up when they would go into a footwear store. I want my footwear to speak to them intuitively. I want it to look like the creation of an individual artisan.

Stern: When you decided to go into business for yourself, you obviously had all the technical skills, because you worked in the shoe industry. But you didn't necessarily have the business skills. How did you approach that when you decided to start a company?

Keen: Well, I learned as much as I could by reading, but you're right: I lacked the business skills. Most designers, myself included, are much more right-brained and do not necessarily revel in the minutiae of spreadsheets and business planning. So I sought a partner who could lend that half of the brain that I didn't have and didn't want to have. I realized that if I put my effort into creating the whole business by myself, then I would not be able to be the creative genius behind the brand. And that's really what makes a brand. It's all about the product.

Stern: Before you decided to launch your own product under your own brand, did you go around to the industry with your ideas and try to market or license those?

Keen: A lot of companies have preset ways of doing things and a certain aesthetic, and I didn't want my idea to get bastardized. I did have a meeting with one company that made me realize that I needed to create this thing myself.

Stern: So at that point, you hadn't decided to launch your business. You still allowed the possibility of having someone else's name on it.

Keen: Yes. I was considering licensing it as a standalone brand. I wasn't considering selling the idea to Timberland or Nike and letting it get absorbed into their brand. I wanted to create something to stand alone that had values that were different from any existing brand out there.

Stern: So you found a partner. You incorporated and you launched the brand of Keen Footwear. Did you feel that you had as much control of everything as you anticipated?

Keen: Yes, initially I definitely did. After we brought a couple of partners on board, there was a struggle for power with my money partners.

Stern: So the new partners were financial partners?

Keen: Yes. Financial and strategic partners. I realized I had to make some sacrifice, even though this was my child. Although I had birthed it, I realized I was going to have to cede some control.

Stern: Was it design control?

Keen: No, it wasn't design control. Being a competitive sailor, I wanted to keep this brand focused around a core product. But I realized early on that the market for sailing shoes sold to competitive sailors would be too niche, and that the outdoor space is much broader. So as far as expanding into other categories within the realm of outdoor footwear, I had no issue. I designed all sorts of product for all sorts of companies over the years, so I didn't have any issues with it.

Stern: So as the company was growing, you obviously became involved on the business marketing side, or at least more keenly aware of it. Do you feel that your involvement in that side changed your design direction? Did you learn anything from it, or was it just a conflict?

Keen: We got into more categories than I had envisioned, but they weren't negative. They were all positive directions. We did bring in a couple of outside designers to help, because we were expanding so quickly. And I couldn't handle all of the design that needed to take place. I happily started working with other designers. So no—there were no issues with that. It was great to be able to expand and do a variety of categories: cycling product, beach product, and a little more women's street fashion product, which was not my forte. We brought in some female designers to support that. It was great to watch the brand expand.

Stern: In participating in running this business, do you think it made you a better designer?

Keen: I would say it did. In the past, I had been working as a pretty independent character within companies. To become design director of a number of designers within various categories definitely made me a better designer, as well as a better manager of my own time. Watching how other designers work has always been enjoyable.

Stern: Do you have an involvement in the business now?

Keen: I sold the company two years ago to my partner. I decided to get out of the footwear business for a while and see what it was like in the furniture world. I had actually come up with an idea for furniture in 1994, well before Keen, and had been designing and building prototypes for my own personal use—not really thinking, "Oh, one day I'm going to go into furniture."

Most good ideas are born of need, and I designed this product while I was designing footwear. I found that I was much more creative and comfortable when I worked in an upright posture, rather than sedentary. I am a very active worker anyway and always moving to and fro. I tried a standing desk for a while, then I tried an upright high stool at the standing desk, and then the aha! moment came in '94.

My high stool was built from an old farm tractor seat and had a little metal backrest on it. One day I was sitting at my desk, hunched over, uncomfortable, and realized: I don't want to stand. I don't want to sit. I want to do something in between. So I bent the backrest of this high tractor seat stool back and I tipped it forward on its front two legs and just started leaning against the stool, so that I was neither sitting nor standing, but very comfortably leaning. Whenever you are at a cookout at somebody's house, you always find yourself leaning against a picnic table or against the fender of a car: it's one of the most comfortable positions to be in. Over the years, I built twenty different prototypes. Two years ago I decided to launch the idea.

Stern: When you sold Keen Footwear, did you feel that you had accomplished your post-9/11 goals?

Keen: Absolutely. I left the industry having made a name for myself.

Stern: What about leaving something that had your name on it and you see a lot of people wearing?

Keen: For me, this is just another brand. I created it, but didn't want to be too closely identified with it. I never wanted this to be a Tommy Hilfiger or a Kenneth Cole. It was never about me, this famous designer, wanting to have my name all over people's bodies. I was not into labels. I have always been a very utilitarian person, just wanting to create a product that worked well. It is an honest product that functions well. It just happened to have a name on it that's in the dictionary and is also my last name.

Stern: When you sold the company, you signed the papers on Friday. When you woke up Monday, was it was just another day?

Keen: Yes. I like to be very active and working. I'm not one to sit back and say, "All right, I'm going to take a year or two years off and see what life's all about." I know what I want to do, which is just to continue to create things and make a difference in the world.

Stern: Do you have multiple projects that you are working on simultaneously?

Keen: Always. I've built prototypes for all kinds of things in my studio. They're sitting on the shelf, waiting for the right time. I like to concentrate on one thing at a time. You achieve the best and most successful product if you're really putting one hundred percent of your effort into one product and one brand at a time.

Stern: In doing any of these designs or inventions, have you ever been confident that you were going in the right direction, but then the product proved to be a failure when it hit the market?

Keen: I don't want to say that I haven't had any failures, but we've not had any major catastrophes. We had a couple of items at Keen that did not achieve the level of sales we expected, but not because our market analysis or designs were faulty. The fault lay with the production quality in the factory.

Stern: So are you now back to working by yourself?

Keen: I am, actually. I hired an engineering firm to work up this product in CAD form and do some preliminary testing prior to opening any tooling. And I hired a design firm up in Pawtucket, Rhode Island, to develop and refine the product. It's really been an enjoyable process working with that team: five Rhode Island School of Design grads and a Rochester Institute of Technology engineer.

Stern: Do you plan to bring this to market in a similar fashion as Keen, in the sense of bringing in a partner and bringing in financial people? The same model?

Keen: To a degree. With Keen I did not have control. My partner basically funded the entire thing. I put in very little money and was a minority shareholder. This go-round, I am funding the whole thing initially, but I'll bring in investors as we go, and I'm not averse to giving away some ownership.

Stern: How does it feel to have that risk? That financial risk this time?

Keen: It's definitely different. I am fortunate enough to be in a position where I made a fair amount of money from the sale of the company. That is a certain amount of risk that you are comfortable putting up as a percentage of what you have. Sure, it definitely makes you think twice about all the decisions when you are opening different tools and committing to certain trade shows and building a booth. The numbers add up, but it's all part of doing business.

Stern: Are you bringing in someone to run the business, as with Keen?

Keen: Yes. In fact, today I've got an operations officer I'm meeting with. We are putting together a team in the next two weeks.

Stern: So your skill set is still not the business side?

Keen: It's visualizing the direction and strategy of the company, and then the business side sort of falls into place around that.

Stern: What business skill sets do you think an inventor needs?

Keen: An inventor needs to see how a business is run and what makes a successful business. Again, a lot of that all comes from products. But you have not just to create a great product—you also have to sell the product, deliver the product, and get paid for the product. So that's sort of the next step. It's that act of actually following through, getting the product manufactured, shipped, and sold to the right places at the right price, while creating the value of the brand and maintaining the standard of the brand and the integrity of the product.

The Internet has radically changed the business side in the past fifteen years. Consumers have the opportunity to customize their relationships with brands and engage them interactively and socially. Particularly as early adopters, they can justifiably feel that they are helping to create and craft a message of the brand. I like to get involved with the consumer at a very early stage and really understand why they are adopting my product. How are they using it? Are they sufficiently strong advocates that they are recommending the product to their friends?

Stern: Do you see the consumer as a partner then?

Keen: Absolutely. I've always felt that. That's why early on with Keen, we started giving away a great deal of profit to charities—just because it was the right thing to do for me. I felt we didn't need to advertise with Keen early on. It actually came when the tsunami hit in Indonesia, and I realized that I've always wanted to try and do the right thing from a social standpoint. It wasn't a marketing gimmick. But when you do the right thing, it doesn't go unnoticed that your company has more social integrity than a company that's just trying to make a profit.

Stern: Along the way, have you had mentors and, if so, how did you find them and what do you get from them?

Keen: My first mentor was Reinhart Butter—one of my Ohio State University professors I mentioned earlier. I admired his German methodology and I really felt that rubbed off on me a great deal. The Germans go about getting the right semantic feel and flavor of a product in a very different way from a lot of American designers. They look at absolutely every aspect of material, form, size, touch, and feel.

Another mentor whom I didn't meet until I was just about to launch Keen in 2002 was Steve Jobs. When I was flying through San Francisco airport, I was sitting there working on my Mac and a guy comes up to me and he's like, "Hey, what do you use your Mac for?" And I say that I am putting together a footwear company and I'm laying out the catalog, and he's like, "Oh, here's the latest operating system." You know, it wasn't even out yet. And I'm like, "Oh, you must work for Apple." And he said, "Yeah, yeah."

Stern: He handed you a CD with the OS.

Keen: Yeah. And he said, "I'm the head of marketing at Apple."

I said, "Oh, do you know Steve? I'd been a Mac user since 1986."

And he said, "Yeah."

I said, "Do you know his shoe size? I want to get him a pair of shoes when I get the molds open."

He said, "Well, he's here in the lounge with me. Would you like to meet him?" This was when he was still flying commercial over the Pacific. So I was fortunate enough to meet Steve, and I got his shoe size, and he was an early adopter. He was one of the first to get the Keen product. I don't know if you recall ever seeing him at some of the product intros: he would be wearing either New Balance or Keen—the product that I sent him. I'd always looked to him as a mentor, even though I never worked with him.

I had that very brief meeting with him and we had communicated a few times since then. But I always appreciated his drive and his absolute perfectionism. I'm nowhere near as anal as he is, but I appreciate that there's one way to do things and it's the right way. I'm definitely not as much of an asshole as he was to the people he worked with. I knew he was very difficult to work with, but I would definitely say he has been a mentor to me in a way.

Stern: Do people come to you now to be mentored?

Keen: Yes. Because I'm close to the Rhode Island School of Design, I spend a lot of time working with an entrepreneurial group there mentoring kids. I was asked by my university, Ohio State, to come there as a distinguished alumnus this past spring, which was great fun. I really enjoy working with students who have good ideas and ambition, and letting them know that it's really not that hard: you just have to have the stick-to-it-iveness that can get you from the ideation stage to going out and sourcing the right people to help you make something real with your dream.

Stern: Do you have any particular products that you like that you have sitting around your studio or on your desk, unrelated to footwear?

Keen: Yeah. Sitting around in my studio are a lot of biological forms that have evolved over the millennia, from shells to bones. I have a friend who is a marine biologist here in Jamestown, and whenever a whale strands on the coastline, he's the one that's called to bring his chain saw and cut the beast up. I've gotten all these bones from him whose forms are deeply inspiring to me as a designer. What could be more functional than the design of bone that evolved to support an animal structurally? Philippe Starck and Raymond Loewy, whose work I find fascinating, consciously apply natural forms, ranging in scale from human to molecular, to their design products.

Stern: Any thoughts on intellectual property?

Keen: I think it's obvious you've got to protect yourself. This is a dog-eat-dog world, and you've got to spend the money to secure ownership of your idea. Even though you know you created the idea, you have to document the fact. There can't be any hesitation to spend money on trademarks and patents and their maintenance. And you should go for a utility patent as well as a design patent. A utility patent is the be-all and end-all. But a lot of things involve prior art, so unfortunately sometimes all you can get is a design patent—which doesn't do a great deal, but at least that's baked in the cake, as it were.

Stern: When you're at a dinner party, what do you tell people you do?

Keen: I tell people I'm an innovator. A brand builder. I like to create something out of nothing.

Stern: What do you do for fun?

Keen: I still race sailboats competitively. I play tennis. I go for bike rides with my wife. I have a couple of black labs that I love to death. So I enjoy taking them down to the beach.

Stern: Do you plan to retire, and, if so, what would you do?

Keen: I have no plans for that, no.

Stern: So will it just be another project?

Keen: Yeah, I think if I were to go into a mode that would be called retirement, it would be spending a lot more time sailing and enjoying Mother Nature. Maybe sailing around the world.

Stern: Any final thoughts or suggestions for an inventor?

Keen: An inventor or designer needs to be able to slow down and observe his own mind with perfect awareness and clarity, free of the fog and clutter and crazy speed that life throws at us. The designer needs to appreciate that while we don't immediately apprehend all the answers, they are there in front of us, and we just need to be able to see them. I'm not a God-fearing person, but I believe that we have a spiritual capacity that we can access by slowing down and observing the moment. Yeah, that's really it. To me, creativity comes from having that space.

Kevin Deppermann

Chief Engineer, Senior Fellow
Monsanto

Kevin Deppermann joined Monsanto in 1978 as an electronic technician after graduating from Ranken Technical Institute in St. Louis. Prior to that, Deppermann served in the US Army, working on electronics in the fixed secure voice communications and data security center. He is now the chief engineer and a Monsanto Senior Fellow leading the Chemistry Crop Analytics & Automation and Engineering Team. In over 33 years with Monsanto, he has developed new innovative instrument processes and machines, which have accelerated Monsanto's breeding, biotech, chemistry, and manufacturing pipelines. Examples of these are the development of large-scale MRI oil-analysis processes; seed-chipping systems for soy, corn, cotton, melon, and wheat; and associated automation with the scalability to transform future plant breeding.

Deppermann holds a BSEE from the University of Missouri–St. Louis/Washington University Joint Engineering Program. He holds certificates in technology management and engineering management from the California Institute of Technology. He has many patents in the areas of automation and engineering. He has also received several Monsanto Above and Beyond awards. He is the recipient of the Monsanto 2009 Science & Technology Career Award, 2010 BAMSL Inventor of the Year Award, the Monsanto 2011 Edgar M. Queeny Award, and the 2012 James B. Eads Award for Outstanding Scientist. Deppermann was also named a Fellow of the Academy of Science–St. Louis.

Brett Stern: Where were you born? And what was your education and field of study?

Kevin Deppermann: I was born in Washington, Missouri. My background, basically, is that I was in the military—in electronics. I then went back to school, got a two-year computer technology degree from Ranken Technical Institute, worked in Monsanto for about ten years, then went back to school and got an engineering degree from the University of Missouri at St. Louis/Washington University Joint Engineering Program.

Stern: Even though you're working with seed development, your background is originally in engineering, not biology?

Deppermann: That is correct.

Stern: When you were growing up, did you consider yourself an inventor? Did you have any inventions?

Deppermann: Well, my mom might have thought I had a bunch of inventions. I was trying to invent how to stay home and get in trouble. My dad grew up on a farm, and because there was always the money-versus-getting-it-done scenario—it was always "just figure it out."

There's always different ways of putting stuff together. I look at everything now, even if my wife gets on me about parts and stuff. If I have a broken thing, I want to save the screws and mechanisms so I can use them to fix something else. I did a lot of fixing, a lot of messing around with radios and car stereos, and cars in general. I messed around with a lot of chemistry for a little while. Erector sets and electronics when I was in grade school and high school.

Stern: So growing up on the farm, I guess you'd sort of call that Yankee ingenuity.

Deppermann: Oh, I didn't grow up on a farm. No, I grew up in town, but my parents had grown up on a farm and my grandfather still had a farm, so I could go out there and do the work. I learned how to milk cows. I wish I could figure out how to use a cherry picker. That was the worst thing in the world—picking cherries when I was a kid, going out there all the time. It takes a long time to pick a gallon of cherries, let me tell you.

Stern: Can you define the technology and field of study you work in?

Deppermann: We do a lot of seed-based and field-based automation when it comes to agriculture. We have developed high-speed pick-and-place automation equipment, so as to read the DNA of a seed before it is grown. We accomplish this task by shaving off a little piece of the seed for analysis so we can still keep the seed. The seed is still viable and plantable. Then we do the DNA analysis, testing that seed for specific traits or things we want to do with those specific seeds.

We have also developed a lot of field automation equipment for testing when it comes to planting. We are trying to determine how to combine positive seed traits and better methods to grow plants from those seeds. It's the whole food chain of automation with respect to agriculture. We do a lot of planting and harvest analysis, and then we go back to planting again. We do anything we can to make the breeding cycle shorter and more efficient. Basically, we are trying to optimize the farmer's yearly crop yield.

Stern: In developing products for the farmers, are you developing machinery? Or is it the actual seeds that you're developing for them?

Deppermann: We don't develop much machinery for the farmer. Most of it is used internally to optimize our processes. We actually help develop the seed that we sell to the farmer.

Stern: How much input do you have from the farmer regarding the final product, the seeds?

Deppermann: Monsanto works closely with the farmers with respect to the final product because we're always looking for ways to make our seeds better, more viable, and more consistent as a function of yield versus time. If we had the perfect seed to plant, we wouldn't need weed control, we wouldn't need anything. It would yield a lot, and then the farmer wouldn't have to worry about weather, wouldn't have to worry about other things. But, that's not the case right now.

Stern: How close is Monsanto to the perfect seed?

Deppermann: I'd say we're quite a ways off. We've got some awfully good stuff now, but it can always be better. One of our goals is to be able to double our crop yields by the year 2030 in the three measured crops—that's corn, cotton, and soy—using one-third less resources. That's pretty big. That would be like doubling your gas mileage using one-third less gas.

Stern: If you stepped back to ten years ago or twenty years ago and had the same goal, would you say the technology has come to fruition?

Deppermann: Well, actually it's kind of interesting you say that because I can go outside my office and look at three fields. The corn production in 2010 was 154.3 bushels, a doubling over the past forty years. What we want to do by 2030 is get to three hundred bushels per acre. We can do that with density. We can work on density and new hybrids to try to get that, but what we're trying to do is to maximize yield per unit area.

Stern: What would you say was the state of the art prior to your investigation in this field of study?

Deppermann: It took a lot longer because you'd actually have to plant the seed, grow them up, and either look at them manually or go out with a paper punch and get a piece of the leaf off each plant. Then you'd stick it in a ninety-six well plate or a tube for analysis, bring it back to the lab, and do a DNA analysis and get some specific traits that you wanted. Then you'd go plant it.

With seed chipping, you can actually do it before you plant, which is a huge thing in respect to process optimization. So, you actually know what you're going to get, the characteristics you're going to plant before you plant it, and you don't have to plant as many seeds.

Stern: So before this, you had to go through a whole growing season?

Deppermann: That's correct. Basically, you harvested the seeds, you got it all intact, and then you planted it. As soon as you had the data, you could theoretically go plant it if in a growing season.

Stern: Did this project get assigned to you or was it something where you had the vision, and you went back to your management and said you wanted to work in this area?

Deppermann: It's kind of interesting because usually in the corporate environment, you try to solve certain problems. My team and I developed this because there was a need for [automation] to be done at high speeds. Most of the time, there are two different ways to innovate. One of them is to spot a need that somehow nobody saw before, and you try to solve it. The other is that a need shows up and you really need to solve this problem.

Stern: In the work that you're doing right now, was it more driven by the latter in that it was something that you saw as the problem?

Deppermann: Yes. In Monsanto, I saw the problem of taking a long time to read the seed DNA, and then figured we're going to have to solve this somehow. As we add more and more traits into the seeds, we had to be able read the DNA quickly or we'd have huge numbers of people out in the field testing each plant. Keeping everything straight would have been a huge undertaking.

Stern: Could you talk about how you work with your team?

Deppermann: I have a team [with people from] a lot of different disciplines that report to me. We have probably twenty-five to thirty people, and then we have a bunch of contractors too, so the team's a lot larger than that. The disciplines are plant genetics and electrical, mechanical, chemical, and software engineering. On the software engineering staff, we have electrical–mechanical technicians, chemists, chemical engineers, and biologists.

Stern: Are you the coordinator? Or do you bring creative and technical expertise as well?

Deppermann: Both. It's my job to help them through things, but also throw challenges at them and then try to figure out what the good themes for a specific problem would be. A lot of times, it's different disciplines and multiple themes working on a problem. It's never just putting one engineer here and one engineer there. It's usually a biologist, a chemist, maybe a couple of mechanical engineers, an electrical engineer, and a couple of software people, and maybe some statisticians thrown in.

Stern: Do you have any particular method or process that you use to get them to work together and motivate them?

Deppermann: Well, the motivation is that I say, "Okay, here's the goal you guys need to figure out." Usually, I just say that it's your job to figure it out. Figure out how you're going to come up with the solution.

Stern: It's somewhat hands-off?

Deppermann: Yes. The reason why I like hands-off is because if I sit there and tell them, then it's just my idea. They need to come together as a group and formulate their ideas. It's not mine unless they really get lost in the weeds or whatever. That's what it's all about—it's about unleashing their creativity.

Stern: When they come back to you with, say, a half dozen solutions or possibilities, what process do you go through to sort those out or show the efficacy of each process or each solution?

Deppermann: A lot of times, we can do quick experiments. Sometimes it's a lot more involved. Sometimes it's a gut feeling about what you think is going to work the best. There's a multitude of processes. It depends on how fast we have to get the solution. If we can go from point A to point B and it makes sense, and we can get it done quickly, we do that. Otherwise, you just nurture the team and say, "Hey, maybe look at this direction and look at that direction." But in the end, they're usually pretty innovative.

Stern: Do you have a particular definition of what innovation is?

Deppermann: That's a good question. I used to get a kick out of it when people would come to me and say, "I only want to work on the hot projects." Well, the problem is I don't know what the hot projects are.

Someone could walk down the hallway and come up with this idea or that idea. You have to nurture these environments. And just because something was tried before—I don't care how many times it was tried—we're in a different time and a different space, and there are different people working at it. So, just because we tried something before and failed doesn't mean we're not going to try it again. I think that's part of what innovation is about. It's about being wrong about ninety-nine percent of the time and then right one percent, and that's what fuels everybody.

Stern: How do you get up every morning knowing that you're going to be wrong ninety-nine percent of the time? What is the motivation behind it? Do you have different versions of motivation that you offer to your workers?

Deppermann: One thing I have is a real short-term memory. It's probably the best if something fails. I just try something else. One of my methods is that I don't really have any fear, and that I'm going to die trying to do whatever I need to do to get it done. Sometimes you just have to put yourself in a different space because you don't want to be bound by things you can't do. I'm a glass half-full person, and most of the people who work for me are pretty glass half full. In fact, sometimes it's even hard to get us back in the box.

Stern: Do you get to pick the people you are going to work with?

Deppermann: No, it's like everything else. This team is the one you're dealt with and you try to get the maximum out of them. You give them as much encouragement and help as you can to get to the solution.

Stern: You just said that you lacked the fear of making mistakes. Would you say that translates into your personal life or your daily routine at your home setting?

Deppermann: That's an interesting question, because my wife would say I'm somewhat hesitant on certain things and that I lack common sense. I don't believe that. She just has different definition of common sense. I'll take anything apart at home. I don't think there's anything in the world that I haven't taken apart and tried to fix or tried to save, except personal problems. That's a whole different story. That seems to be a whole new domain of innovation, which I haven't quite figured out. Especially with my kids.

Stern: You seem to wear two hats. One as a technical person, where you seem to thrive, and the other as the manager of all these other people running around. How are you the doer, but yet the daddy at the same time?

Deppermann: I think part of it is that you have to try to do things for other people. Show them how to fail and then move forward. In a lot of ways, the new engineers, biologists, and chemists coming up don't understand failure because they've never really had it in school. They don't know what not to do. In some ways, I don't know what not to do to, but I'm smarter picking out what to do. But I don't really take failure as a bad thing.

Stern: Do you have a different term for it then?

Deppermann: Let's see. Oh, it's just *experiment*. Experimenting is nothing but failure.

Stern: When you go back every quarter and look at your budgets, how do you explain to your management, the people overseeing you and your budgets, that you're just spending your time failing or spending your time experimenting? How do you justify that?

Deppermann: You know, we still have base technologies that we can grow on. But you need inventions along the way, or at least new thoughts along the way, to be able to invent something—new processes, stuff like that. You've just got to be right more than you're wrong.

Stern: When you're working on the technical side, do you like working alone or in a group?

Deppermann: Either one. I used to spend a lot of time driving back and forth from St. Louis to Iowa. The drive would usually take five to six hours to get up there and then five to six hours to get back. A lot of times, I wouldn't even turn the radio on. It's amazing how much time I think about just imagining myself as a plant in the field. How would you generate more yield? What would you need? How would you do it? It's all about envisioning the outcome even though it may be wrong. But I'm always looking at the outcome versus the steps involved.

Stern: While you're driving along these open fields, you're picturing yourself as a plant, as a living seedling?

Deppermann: Yes.

Stern: As the sun coming up, as the water coming down, and going through that motion?

Deppermann: Yeah. The process where I put myself in there, I do that a lot.

Stern: So, you really enjoy the quiet time?

Deppermann: Once in a while—because it's not that quiet here. It's pretty hectic most of the time. There's always something to be done. Quiet time does give one time to reflect on things and try to put pieces together or maybe even think about new ways of doing things.

Stern: How you define a problem or ideate? Do you have a particular process?

Deppermann: That's a good question. Sometimes I can see a problem, the definition, pretty fast. But there are multiple components to it. Let's say a manufacturing process for some reason went awry. You have all these variables you have to deal with and try to figure out, but it's pretty easy if you can define the beginning and the outcome. Everything else is a journey in between.

Stern: So how you get there is wide open?

Deppermann: Oh, absolutely. You know, at the end of the day, how we get there is the biggest thing. The first is being able to do it, and then we'll figure out how to take the cost down or make it more efficient. I always say, if you can't do it manually, you're never going to be able to automate it. But I've been kind of changing my tune a little bit because some of these problems you can't do manually. They can only be viewed in virtual space.

Stern: Is the technology just becoming too complex or too automated?

Deppermann: I think in a lot of ways the technology is becoming more complex, but the understanding of the technology is less and less known. I see that a little bit with respect to having all these whiz-bang tools nowadays. The amount of time you actually spend on how they were generated versus the fundamentals of what you're trying to do. I have some engineers who can do all kinds of stuff in 3D space, but there's a big difference between looking at it on a computer screen and actually making it work in real life.

Stern: Do you consider yourself an engineer or an inventor?

Deppermann: Both.

Stern: When you go to a dinner party and someone asks you what you do, what is your response?

Deppermann: Well, the first thing is usually, "I lead an engineering team and we're responsible for doing innovation. I think they go hand-in-hand."

I think the best compliment I ever got from my boss was one day when he said to me on an airplane, "Somebody asked me what you are. You're not like any engineer I've seen because most engineers need their input specs—step one, step two, step three—to get there." But, I'm a lot different. I don't do it that way. I don't care if I need a step. I'll go out and figure out one. Or maybe there are all kinds of steps. Maybe I'll have four or five scenarios in my mind at one time to try to figure out how to do something.

Stern: What particular skill set do you have and/or what does an inventor need to be innovative?

Deppermann: I think an inventor needs a bunch of skill sets. You have your fundamental education. But a lot of times, it's a dogged determination not to be wrong, or to make sure you can solve the problem. I think the worst thing anybody can tell me is that I can't do something. Let me tell you—I'll figure out a way to do it.

Stern: Is it to spite them? Or to just prove it?

Deppermann: Probably both. I think in a lot of ways, when you get into a subject that you don't know, you're asking questions and you're not bound by any prior knowledge. You're not bound by any preconceived notions of what you can't do.

Stern: Could you tell me about some of your biggest failures?

Deppermann: I've had some robotic processing problems that were failures. But they were basically because I think a lot of failures are about people trying to get people to work together to find a solution in a specified period of time. I think that's getting harder and harder to do. I think that trying to line everybody up in the right direction and thinking about it in the same way can be very cumbersome. I think the biggest thing is what you learn from failure.

Stern: Let's talk about the work that you do personally. When you ideate or brainstorm, do you have any particular methods? Do you like to sketch? Do you like to prototype? What are the tools that you use to for brainstorming ideas?

Deppermann: Most of the time, it's just sketching or throwing ideas out. I can sketch up something pretty fast with respect to see if it works. Or I just go into the lab and throw some stuff together. I don't need expensive instruments. Just some metal and some electronics or whatever, or even just looking at it from a whole different point of view. It's one of the reasons at home I just get some random parts that are lying around and throw something together and see how it works.

Stern: So, you're proverbial sketch-on-the-napkin type of guy?

Deppermann: Yeah, believe me, I've done a lot of things on the back of a napkin. A lot.

Stern: Do you have a pile of napkins just sitting around?

Deppermann: Well, they do go from napkins onto paper and then to a CAD system as a function of "Let's try this, let's try that. Well, that didn't work— let's try something else."

Stern: But sketching is important to you?

Deppermann: Sketching is very important. I don't want to go to all the work of putting it in the computer. I'll just sketch it in two minutes.

Stern: Does your team use that same process, or are there different methods?

Deppermann: Oh, a lot of them have a method where they are writing in text what they're trying to accomplish. Some of them just go to CRT screens or CAD packages and run it up. I guess maybe I'm old school, because I really like a pencil. It's just the easiest way for me to do it. Plus, I don't need to learn a whole lot of packages to be able to sketch well.

Stern: You just have to sharpen the pencil.

Deppermann: I can track decimal points pretty easy in my head, so just give me a few numbers. I'm pretty good at that. I can figure out pretty fast if it's to scale, or it's the right quantity, or the magnitude of the problem or whatever. I think that's a pretty important skill. Sometimes you have to have a sense of whether it's going to work or not because, like I said, sometimes it's just a gut feeling.

Stern: You obviously trust your judgment.

Deppermann: Most of the time.

Stern: At what point in your career did you feel this trust instinct? Is this sort of a maturity that happens, or through enough experience or failures or successes?

Deppermann: I think I kind of always had that because my attitude was always like, "Even if this doesn't work, I'm going to try another way." There was never failure. Failure was never an option for me. I may have failed on certain things, pieces of it, but I always had another idea. I've never run out ideas to pick or try something different.

Stern: Generally, what are the timelines for a project that you work on? Are they six months, a year, five to ten years?

Deppermann: They can go anywhere from a week to as long as a multiyear project. We've got some projects that have lasted four to five years. We have some technologies that we've been working on for the last five, six, seven years. We've made a lot of breakthroughs, and then we need to make more incremental breakthroughs as we go. A lot of the time, it's just, "Can we do this at all?" And then if you can, after the first reiteration, then you're on your second reiteration, and then third reiteration. Seems like when you solve one problem, you then create another one.

Stern: How many projects do you work on simultaneously?

Deppermann: The team has anywhere from seventy-five to one hundred projects at any time. I kind of believe that a good inventor who is pretty innovative should be working on multiple projects simultaneously because when one gets a little bit taxing, you go to another one.

A lot of time what happens if you work on one project over and over, you kind of get narrowed with your scope on what you want to file. So I really think it's important to have a diversity of projects that you work on simultaneously.

The funny thing about engineering is that a lot of times, engineers keep working on a problem, and they get a little bit more narrow-minded about where it's going as they get farther and farther into the project. It's kind of like if you drive the same way every day, you get in a habit of just driving. So you need to have something to take you out of the mode for a while, maybe a diverse project, something totally unrelated to anything else you're doing. It gives you more insight, more information—just new information in a different way that can help with another project. It's amazing.

Stern: In the technology you're working on, and potentially with Monsanto, what do you see as the future for the next generation of the technology?

Deppermann: That's a good question. Well, hopefully it will lead to the perfect plant, whatever that is. There will be a merging of engineering, science, biology, and a lot of different things to get there. The interesting thing about this is, with seeds, you're constantly dealing with the organism, wondering how much you can deal with it and still make sure the plant is viable for planting.

Stern: Could you talk about the commercialization side of the process? Do you get involved at all in that—in the selling or the marketing of it?

Deppermann: Most of what we do is research. It's all with respect in trying to produce better seeds, better methodologies, and better growing systems for our plants. A lot of that is about coming up with ideas and advancements to try to help that along. But advancements need to be carried out by the commercial teams.

Stern: You seem driven by the motivation to solve a problem.

Deppermann: Absolutely. That is number one.

Stern: Does financial reward ever drive your motivation at all?

Deppermann: Not one bit. Not one iota. I make enough money that it's not about the money. It's about the problem. If that ever changes, then I won't be that interested in doing it. In fact, after I got over the initial thing where you've got enough money for your family and all that, I didn't even look at my pay stub anymore. It's not what drives me. What gets me up in the morning is the challenge of the day. There will be a new challenge, whether it's trying to motivate my team, or a new project, or trying to operate in the chaos of invention. That's what it's all about.

Stern: Do you have any professional heroes? People that you look up to?

Deppermann: I think my number one is Steve Jobs. That should be pretty self-explanatory. Thomas Edison is one, and Richard Feynman, the physicist. It's pretty interesting to see how people got to where they are. But also understanding what their thought processes were when they had their backs against the wall, trying to solve something.

And if you look at the ones that are really successful, they failed a lot. I got a kick out of reading something about Abraham Lincoln and all the failures he had before he became president. It was pretty amazing. I read it to my team and none of them got who it was. He failed in business and [the message was simple]—you have to pick yourself up and have the self-confidence to go on. In your career, you're going to have ups and downs, but it's just the matter of how you handle them.

Stern: Are there any particular inventions that you admire out there? Any products that you really like?

Deppermann: I think whoever invented the remote control for the TV was a genius. I don't really have anything that amazes me that much except for the remote control.

You know, my grandfather was born in 1880—and look at all the inventions he saw. He saw the airplane, cars, and tractors. He went through wars. I mean, holy cow, you look at some of the inventions that he saw. Will my generation ever see that much? When I was growing up, you didn't have computers. I remember when the first calculator came out—after having a slide ruler.

Stern: Besides your pencil and your pad, are there technologies that you like using in your daily routine that you couldn't live without?

Deppermann: The calculator. I have an HP calculator. That's like the best thing ever invented.

Stern: If you can talk about it, is there a next project down the road or something that you want to work on?

Deppermann: I think it would be cool—it might not happen in my lifetime—but I think the *Star Trek* transporter would be one if somebody could figure that out.

And it's not unusual if you think today. Look at transportation of information—it used to be letters. You don't need those any more. It now miraculously shows up. A lot of the engineers we have here design stuff in their 3D terminals, it gets sent out to a manufacturer somewhere, and suddenly it's a [physical] part that gets shipped back to us. So, we're already doing some of that, but not at the level that we need to make transporters real. We're in the infancy of that.

Stern: Do you have any advice for would-be inventors out there?

Deppermann: Well, the only advice I have is dogged determination, to not be dissuaded. In other words, die trying and no fear.

Stern: What do you do for fun or distraction?

Deppermann: I think the most fun I have is just messing around. I really don't have anything per se as a hobby. I like to work on cars. I really think some of the technology in cars is truly amazing. If you look at some of the airplanes that were made back in the thirties and forties during the war, they are just incredible—with the tools they had and how they made them. They didn't have all the fancy stuff we have now. It all had to be hand built. I have a lot of respect for craftsmanship. A tremendous amount of respect for craftsmanship.

Stern: Do you plan to retire?

Deppermann: Maybe when I'm dead.

Stern: You're just going to keep working until they kick you out the door?

Deppermann: That's probably true because, until I get uninterested in what I'm doing, which should be a long time, it's all about challenge. If we're not here to change the world or make the world a better place, then why are we even here?

Stern: Would you say the success that you've obtained has changed your life?

Deppermann: It hasn't at all. Not one iota. I think of things exactly the same over the years. The thing is not get too engulfed in success. Don't worry about it because it's just for this period. I really think it's the drive and determination to be successful that's most important.

Stern: I feel your enthusiasm and I'm sure your team really feels that enthusiasm that you bring to a project.

Deppermann: Well, I hope they do. I try hard every day to do it. It is rewarding to see the team take off. The thing that's most important is to never give up, no matter how bad you think the problem is. There's got to be a way to get out of it. Get more information. I think self-resolve or self-motivation is the biggest thing for a person. And it needs to be re-stoked every day.

Stern: That's the skill set an inventor needs? Self-motivation?

Deppermann: Absolutely. I think we need more and more people that are more self-reliant and can actually envision the future. I hope it never gets to be where everything is so cut-and-dry. That there is nothing to invent any more. But I don't think it will ever be. We'll just have different problems.

Stern: We have a different view of the world once these new inventions happen, so there are always continuous problems out there to solve.

Deppermann: That's true. It's amazing how some new inventions create more problems.

John Calvert

Acting Associate Commissioner for
Innovation Development

Elizabeth Dougherty

Director of Inventor Education, Outreach,
and Recognition

Office of Innovation and Development,
United States Patent and Trademark Office

*As the Acting Associate Commissioner in the Office of Innovation Development at the
United States Patent and Trademark Office (USPTO),* **John Calvert** *develops ways
to increase the USPTO's presence in the independent inventor, small business, and
entrepreneurial communities. He earned a BS in textile technology from North Carolina*

State University, worked for 13 years in the textile industry, and joined the USPTO as an examiner specializing in textile technology. He subsequently served as acting director for the Office of Independent Inventor Programs, senior advisor for the Office of Innovation Development, and the administrator for the Inventor Assistance Program. Calvert received the Department of Commerce Bronze Medal and the United States Patent and Trademark Office Exceptional Career Award.

As the Director of Inventor Education, Outreach, and Recognition in the Office of Innovation Development at the USPTO, **Elizabeth Dougherty** develops, implements, and supervises programs that support the independent inventor community, small businesses, entrepreneurs, and intellectual property. She formerly served as acting deputy director in the Office of Patent Legal Administration at the USPTO, and as a patent examiner in the area of electric devices used for measuring or testing. Dougherty received a BS in physics and a JD from the Catholic University of America.

She is a member of the Virginia State Bar, the Giles S. Rich and Pauline Newman American Inns of Court, the American Bar Association, the Federal Circuit Bar Association, the American Intellectual Property Law Association, the Patent and Trademark Office Society, the Supervisory Patent Examiners and Classifiers Organization, Women in Science and Engineering, and the Prince George's County Historical Society.

Brett Stern: What is the function of the Office of Innovation Development within the United States Patent and Trademark Office?

John Calvert: It's to help inventors through the process of working with the US Patent and Trademark Office. It helps them to understand what happens in the patenting process, and tries to get them to the next step once they have this nice piece of paper with a gold seal and a red ribbon on it.

Elizabeth Dougherty: We also work to help demystify the process. To paraphrase an observation made by Mark Reyland, Executive Director of the United Inventors Association [UIA], oftentimes for inventors, as well as general members of the public, the government is "the big scary monster under my bed." Part of the work of the Office of Innovation Development is to put a face on the federal government and make it personable.

Our office aims to make inventors realize that there is assistance here for them as they begin their innovative journey. Once they have that first idea, they want to know what to do with it, what the process is, and how to go about getting a patent. We bring our services to our stakeholders in a variety of fashions—for example, in person, through an independent inventors' conference, through our web site, or through online inventor chats—we make vital information accessible.

Stern: What are some of the common mistakes that inventors make when they first start the process?

Calvert: Thinking that, if they have a good idea, they're going make a million dollars. Most of the inventors that I see have great ideas, but they don't yet have inventions. They don't know where it's going to go. They don't know how to market it. They don't know how it's going to fit on a shelf, or if it is going to be red or blue. And that could be a big sales difference.

Dougherty: Their most common mistake is to fail to do their homework. They have the common misunderstanding that because they haven't seen their invention on a store shelf, it doesn't exist. They haven't done their homework to look and see what has, in fact, been patented before, although it might have never been produced. They assume that because they haven't seen the invention before, the idea must be fresh and new.

Stern: Why do you think becoming an inventor is a quintessentially American ambition?

Dougherty: We've always been problem solvers. Throughout our history, people have come to America looking for the freedom and opportunity to be independent, create something of their own, and claim a piece of the universe. This quest has driven our success as a nation.

Calvert: And individual success stories—Henry Ford, Alexander Graham Bell, and so on—inspire people to think, "Hey! I could do that, too!"

Stern: Can you talk about some of the programs you offer?

Calvert: We have a four-tier telephone-answering program here. People call our 1-800-number and are patched through to the right tier—for example, to the Inventors Assistance Center, which is staffed by retired supervisors and attorneys with professional backgrounds in particular areas of patent process and examination.

Another valuable resource for independent inventors is our Patent and Trademark Resource Centers, which are located all over the country. They used to be called Patent and Trademark Depository Libraries, but we rebranded them. These centers are staffed by specialized librarians who help inventors with patent searches and forms. We also partner with inventor workshops, such as the Tech Shop in the Bay Area, which has an onsite hotline that gives inventors instant connection with the USPTO when questions come up as they're working on their prototypes.

Another important program for independent inventors is just getting off the ground. The Smith-Leahy America Invents Act, enacted in 2011, mandates the USPTO to work with and support intellectual property law associations across the country to create pro bono programs for financially under-resourced independent inventors and small businesses. That mandate falls directly on our office. By the middle of 2014, we plan to have pro bono programs in all areas of the country. Independent inventors who don't have the resources to hire a patent attorney to help them write and prosecute a patent application may qualify for pro bono services.

Dougherty: We don't want the financial barrier of obtaining legal services to preclude someone from bringing their idea to fruition.

Stern: Do many inventors get patents without counsel?

Calvert: We receive between 12,000 and 13,000 patent applications filed without an attorney or agent, or *pro se*. Fewer than thirty percent of those applications go on to receive a patent. Patent attorneys know how to write claims, which is the biggie. In addition, they understand what the Patent Examiner's job is and how to respond to what the Examiner says in an Office action.

Dougherty: Also, patent attorneys are adept at the mechanics of the process; they are knowledgeable of the patent laws and rules and can carefully address deadlines and applicable fees.

Stern: Would you recommend that inventors contact the USPTO before filing their patent applications?

Calvert: Yes. At the very least, they need to acquaint themselves thoroughly with the USPTO web site.

Dougherty: Yes, our web site is the first place to go. We also recommend that prospective applicants consult local inventors' organizations in their area. Learn something from the experience of others who have gone through the process. Those folks can very often recommend a local patent attorney. You can get local contact information from the USPTO web site's Inventors Eye publication[1] or from sources such as the United Inventors Associations web site[2]. These online resources are exciting because they reflect how vibrant the inventive population is.

Calvert: And that community is growing. At one point in time, we saw a downturn in the number of inventors groups. Now, they're growing like crazy.

Stern: Why is that?

Dougherty: The Internet has made information more accessible and facilitates participation in local groups.

Stern: Women and minorities are not proportionally represented in the independent inventors' community. What is your office doing to redress that imbalance?

Dougherty: Historically, those two segments of the population suffered from more limited access to capital and education. Those societal barriers are falling. The USPTO is undertaking specific actions, initiatives, and partnerships

---

[1] www.uspto.gov/inventors/independent/eye/201206/index.jsp
[2] www.uiausa.org

to accelerate that process. Some examples are our work with the Native American Intellectual Property Enterprise Council [NAIPEC], the Society of Hispanic Professional Engineers [SHPE], and the National Society of Black Engineers [NSBE]. By collaborating with these organizations and others we are able to facilitate access by diverse populations to our services.

We have twice annually sponsored the USPTO Women's Entrepreneurship Symposium in partnership with Senator Mary Landrieu of Louisiana, who chairs the Senate Small Business and Entrepreneurship Committee. We fully anticipate continuing this annual outreach event. We also partner with the National Women's Business Council and the organization of Moms in Business in outreach to women and minorities.

Calvert: When I was in college, there were very few women in engineering school.

Dougherty: Try being a woman in physics! In the past, female students were far more likely to go into the chemical disciplines than mechanical or electrical fields.

Calvert: But on college campuses today, I see almost a fifty-fifty women-to-men split in many engineering classes, as well as many more Hispanic, African American, and Asian American students than formerly.

Dougherty: In the executive branch of government, the White House Council on Women and Girls has an intense focus on fostering the engagement of women and girls in STEM [science, technology, engineering, and mathematics]. At the USPTO, three of our four top executives are women: Deputy Director Terry Rea, Commissioner for Trademarks Deborah Cohn, and Commissioner for Patents Peggy Focarino.

We're additionally building a partnership right now with the Association of Women in Science [AWIS]; we're planning to build webinars and a presence to help bring more female engineers and scientists into the USPTO.

Calvert: One of the local inventor organizations in the USPTO's Inventors Eye network is the Chicago First Black Inventors/Entrepreneurs Organization, founded and chaired by Calvin Flowers.

Dougherty: They're an incredibly active inventors' organization.

Stern: What common mistakes do inventors make when they file for patents?

Calvert: Failure to tell the whole truth. They're very secretive. They don't want to tell us everything they know about their invention. They've been keeping everything secret for so long that when they finally get to the point of disclosure, they can't bring themselves to reveal what's in their head. Independent inventors are notorious for this. They won't even confide in their patent attorneys one hundred percent because they're afraid the attorneys will steal it. When I was an examiner and later a Supervisory Patent Examiner and I was in interviews with an inventor and an attorney, I would have the inventor tell me all about his/her invention. I would find myself looking at the patent application and specification and saying, "Oh, wait, you just said something that I don't see in here."

Dougherty: Sometimes they fail to put the best features of the invention in their disclosure.

Stern: Is that just the human trait of secretiveness?

Calvert: I think they think: "Nobody but me has to know everything about this."

Dougherty: Other times, they make the opposite assumption. Inventors know their inventions so well that they often think things are inherent in their disclosure or obvious to everyone else reading that disclosure. So they fail to give the necessary detail to adequately describe their invention as is required by the law.

Calvert: The biggest problem that I see when inventors file *pro se* is jumping the gun. They file before they've finished their homework and figured out what their real claim to novelty is. They haven't figured out how to leverage what's already disclosed in somebody else's invention to their advantage.

Dougherty: Further to John's point, it behooves applicants to know what the relevant marketplace looks like, so they can specify how their device advances beyond current technology.

Calvert: In addition, applicants need to know whether they have what is called Freedom to Operate in the marketplace. Your idea might be novel enough to be patentable yet still run into a "patent thicket" of contiguous patents that would make it hazardous to bring your product to market without licensing other patents.

So, beyond understanding what needs to go in the patent application, an inventor has to chart out the path for bringing the invention to market. Are there other patents already out there in the world that might challenge my product for infringement?

Dougherty: There's also a flip side to inventors being too secretive in their disclosures to the USPTO. Too often inventors disclose too much to the wrong individual or organization when they should be demanding a confidentiality agreement or not disclosing anything at all to them.

Calvert: And when the "first-to-file" provisions of the America Invents Act come into force in 2013, inventors will need to be even more cautious about what they disclose to whom about their inventions.

Stern: What is "innovation" as compared to "invention?"

Calvert: "Invention" is building something new. The next step is taking that something new and getting it to a working model, ready to be commercialized. "Innovation" is that step between invention and commercialization. Where and how am I going to manufacture and market this? What are my costs and margins?

Stern: Would you say patents or other devices for asserting intellectual and commercial property are necessary to commercialize?

Calvert: Not necessarily.

Dougherty: Not always, each situation is different.

Calvert: If you have a product that is going to be in and out of the marketplace quickly, why would you want to spend a lot of money and time on getting a patent?

Dougherty: In other situations, it can be better not to disclose the details of an invention and to keep them as a trade secret. It's a business decision, ultimately.

Stern: Any thoughts on the commercialization side?

Dougherty: We encourage inventors to take advantage of other federal and state resources. We regularly partner with local Small Business Development Centers [SBDCs] and US Export Assistance Centers [USEACs]. We strongly encourage inventors to use these resources as well as the Patent and Trademark Resources Center [PTRC] in their state. Other local places to find assistance with various aspects of commercialization include the Manufacturing Extension Partnership [MEP] facilities of the National Institute of Standards and Technology [NIST] in your area. In addition to these, the Department of Commerce offers many other resources to inventors.

Calvert: Inventors can also collaborate with graduate or undergraduate students at universities, business schools, and even community colleges.

Stern: To what extent do you think financial reward is the driving factor for inventors, as compared to just getting their invention out there?

Dougherty: Financial reward for many is certainly at the forefront of their minds in getting started. As was said previously, oftentimes inventors believe a patent alone is their key to making money.

Calvert: I think once inventors have made their mark with highly successful products, then it sometimes becomes more a matter of reaching for impact and renown. Gary Michelson and Lonnie Johnson, for example, are at the point where they are striving for historic recognition.

Stern: Do you find that most inventors have made working prototypes by the time they get to the patent process? Or is it still on paper?

Dougherty: I find in my interviews with patent applicants that very few of them have actually prototyped. It's usually just something still on the back of a napkin. Depending on the technology area they're inventing in, it may not be something you can make in your garage or at your kitchen table, because you need the access to things such as polymers or processes such as heat manufacturing or other tools that you don't have at home.

Calvert: I agree. I don't see many prototypes. Every once in a while I see one and say, "Wow! What the heck is that?"

Stern: Any thoughts on working with inventing/promotion firms?

Dougherty: Just as with any other kind of investment, whether it's the stock market or purchasing a car, inventors need to do their homework!

Calvert: Right. And if you do use an invention/promotion company, always get a second opinion from someone else and do your own due diligence with the help of a government agency or web site.

Dougherty: One possibility is to get at least an initial hour with a patent attorney.

Stern: Any suggestions on how to find and work with a patent attorney?

Dougherty: I urge people to start with the register of attorneys and agents on the USPTO web site. You definitely want a practitioner who is registered to practice before the US Patent and Trademark Office, and that list is searchable by state and by name. As we suggested before, ask other inventors in your area, too. Go to your local inventors group and ask for recommendations. You need to find an attorney or agent who you are comfortable working with because, in fact, you're developing a relationship. As a patent applicant you're coming to the government asking for a legal right, so you want your attorney or agent to be someone to whom you can fully disclose your invention and in whom you have confidence that they're going to advocate zealously for you.

This might be a great time to mention something that often comes as a surprise to independent inventors: *you have the opportunity to communicate directly with your patent examiner.* While communications with the Office are conducted in writing, you also have the opportunity for telephonic interviews and/or in-person interviews. You can request interviews yourself if you are representing yourself, pro se; but if a registered attorney is representing you, you can still participate. The Office will only communicate with the attorney, to avoid duplicate correspondence, but the inventor has the right to say to their attorney, "I'd like to have an interview, and I'd like to go on that interview with you."

Now, that's a discussion to have with your attorney, and your attorney may advise you against it—but you may be part of the interview process, even with an attorney. An interview is an incredibly effective way to achieve a true meeting of minds, to make sure that your examiner understands your invention and conversely that you, the inventor, understand what your patent examiner has presented in his/her Office action to you.

Calvert: You can actually now request an interview before the examination is done. There the examiner will give you certain information and will pull the prior art and let you take a look at it, so that you can try to get your claims fixed early.

Dougherty: Oftentimes, inventors who are prosecuting their application *pro se* are unaware that they may ask the examiner for assistance in drafting allowable claims if there is allowable subject matter in the written disclosure. The examiner's function is to allow valid patents. So, they will help the inventor come to an allowable subject matter if it exists in the application.

Stern: Which technologies or fields exhibit high-growth trends in terms of patents?

Calvert: One area that is going to be big is bioinformatics, which is biology and computer software working together.

Dougherty: Medical device art is a high-growth area, too. People are living longer and they're seeking to reduce costs for an enhanced life. Devices are getting smaller. Nanotechnology is already enabling medical devices, for example, that can travel through your bloodstream, collecting and reporting medical data in real time.

Calvert: Another area that's booming is electronic games and betting devices in the gambling industry.

Dougherty: For that reason, the USPTO is predominantly recruiting and hiring patent examiners in such areas as electrical technologies, communications, computer systems, and gaming.

Calvert: Also mechanical technologies, especially in the areas of medical devices and manufacturing devices.

Stern: Those technologies are certainly very complex and tend to be more from a corporate side.

Calvert: Not necessarily. We see a number of doctors and dentists who individually invent in the medical technology.

Stern: Do you think it's good to work on multiple projects at a time?

Calvert: I think some inventors need to work on multiple projects at a time. That way they don't get too attached to one item, and they can give up the one that's not been really working out.

Dougherty: I could see that being a challenge for some, though. They could become scattered and lose focus and never bring anything to a final resolution. You have to make a decision. You should not allow yourself to be discouraged by others, but you do need to listen very closely to others. Oftentimes, the inventor is given advice—whether communicated by the patent examiner or by a person who's helping you to try to commercialize a product—indicating that the invention really might not be successful in the market. You have got to be open-minded and objective about the advice you're receiving.

Stern: Any final thoughts on how to deal with the USPTO?

Calvert: I think the advice we give to every inventor is this: don't think that the USPTO is out to get you. We're there to help you. Ask questions.

Dougherty: Yes, avail yourself of USPTO services. Apart from your filing fees and the costs associated with prosecuting your application, the USPTO offers a great deal of assistance for free. Communicating with the Office of Innovation Development, contacting the Inventor's Assistance Center, and all the other resources we've talked about—it's all free help. Make the most of it!

Calvert: And, don't wait until it's so late that you're in real trouble. With any government agency, but especially when dealing with the US Patent and Trademark Office, there are deadlines, and those deadlines are statutory— they're set by law. We can't change them, and you've got to meet those deadlines or you're in real trouble. It's hard to accommodate an applicant who comes in two days before the expiration of his or her six-month statutory period of time to respond, and says, "I've got a problem with my examiner. What can I do?" Well, you should've picked up the phone two days after you got your rejection, called somebody, and said, "I got a problem with my patent examiner. I don't understand what he wants." Or ask your attorney, "What does this mean? What do I have to do?"

Stern: Be timely in your actions?

Dougherty: Yes. If you're going to go to the length of filing a patent application, take it seriously. It's a full-time job prosecuting your patent application—though we understand you're probably doing it in addition to two other jobs. But do take it seriously. Be conscious of the deadlines, and be conscious of the process.

Calvert: If you're conscientious about the deadlines, you'll get through the prosecution quicker, cheaper, and with a much better chance of a successful outcome.

Stern: Any final thoughts that you want to offer?

Calvert: I would urge the first-time inventor to read as much as possible about your invention's market and about the patent process. Avail yourself of your local inventor clubs and Patent and Trademark Resource Center. Go to USPTO independent inventor conferences. It costs a plane ticket, less than $100 for the conference, and your hotel room. You get two days immersed with professionals who know the industry inside out and who are dispensing loads of advice. Do that. Learn. Do everything you can to advance before you even think about filing a patent application. Learn what you need to go through, because it's not cheap or easy to file a patent application. Don't let your invention sit on the shelf. Don't let it languish. File it with the Patent Office and prosecute your application the right way.

Dougherty: It just comes back to doing your homework. Read. Read particularly those patents that already exist in your space, and copy the format of those patents.

Stern: Look at the prior art?

Dougherty: Look at the prior art cited in patents that are pertinent or related to your invention. See how the patent claims are drafted. If you're going to make use of the services of a patent attorney or agent, the more work you can do on your own - up front, the more you're going to save yourself in legal costs, and the better prepared you will be to interact with your attorney or agent.

Calvert: Build a good business plan. Getting a patent is getting a patent. It's not getting a business started. In order to get a business started, you need a business plan.

Dougherty: That's where resources like the Small Business Development Centers are incredibly handy. They offer free or very low-fee classes on how to develop a business plan, as well as classes on intellectual property. They're a great resource and they're located all around the country.

Calvert: These are the things that an inventor really needs to learn and work out before taking that big step of sending a patent application to the USPTO—hopefully as the prelude to starting up their own business.

Steve Wozniak

Co-Founder

Apple Computer

A Silicon Valley icon and philanthropist for more than thirty years, **Steve Wozniak** helped shape the computing industry with his design of Apple's first line of products, the Apple I and II, and influenced the popular Macintosh. In 1976, Wozniak and Steve Jobs founded Apple Computer Inc. with Wozniak's Apple I personal computer. The following year he introduced his Apple II personal computer, featuring a central processing unit, a keyboard, color graphics, and a floppy disk drive. The Apple II was integral to launching the personal computer industry. Wozniak is named sole inventor on the US patent for "microcomputer for use with video display."

In 1981, he returned to the University of California at Berkeley and finished his degree in electrical engineering/computer science. For his achievements at Apple Computer, Wozniak was awarded the National Medal of Technology in 1985 by the president of the United States. In 2000, he was inducted into the Inventors Hall of Fame. He received the Heinz Award for Technology, the Economy and Employment "for single-handedly designing the first personal computer and for then redirecting his lifelong passion for mathematics and electronics toward lighting the fires of excitement for education in grade school students and their teachers."

Through the years, Wozniak has been involved in various business and philanthropic ventures, focusing primarily on computer capabilities in schools and stressing hands-on learning and encouraging student creativity. Making significant investments of his time and resources in education, he adopted the Los Gatos School District, where he provides direct learning experiences and donates state-of-the-art technology equipment. He founded the Electronic Frontier Foundation and was a founding sponsor of the Tech Museum, the Silicon Valley Ballet, and the Children's Discovery Museum of San Jose.

*Wozniak currently serves as chief scientist for Fusion-io. His bestselling autobiography—*iWoz: Computer Geek to Cult Icon: How I Invented the Personal Computer, Co-Founded Apple, and Had Fun Doing It—*was published in 2006 (W. W. Norton). His television appearances include* Kathy Griffin: My Life on the D-List, Dancing with the Stars, *and* The Big Bang Theory.

Brett Stern: You talk about having an engineering side and a human side. Any thoughts on what the difference is, and how you define those sides?

Steve Wozniak: I talk about the difference between the engineering side and the human side in two different senses. One is the general sense of developing technology products. If you look at Apple history, you'll find out that the most important thing that made Apple great—and made Steve Jobs such a great person—was our focus on understanding the users more than understanding the technology. We work and work and work our hardest to try to hide the technology from humans. The result is lovely experiences for the masses and more successful products for us—though maybe not for the geeks in us.

So the other sense of the difference between the engineering side and the human side that I talk about is my personal sense. I'm so thankful for some incredible engineering expertise that I got on my own, just because developing computers was my thing that I loved to do. But in my own development as a person, I have always valued the human side of my work even more —helping and caring for people.

So the human side and the technology side need to be blended and harmonized in the enterprise and in the person. If I have some abilities to build devices. I have to put the extra work and empathy into the technology to make it work in an understandable way to human users. It's got to make sense to them. It's kind of this idea that consumers win in the end. They vote with their pocketbook. They buy what they want. Steve Jobs spoke quite a bit about the importance of understanding what people will like. The thing of first importance is to make the product that people want. The second thing is worrying about everything in the world like patents and different technologies and whatnot. Just do it right. That's the main thing.

Stern: It seems like such an obvious thing. Why do you think it's a challenge for other companies to replicate your emphasis on delighting the user?

Wozniak: You almost have to believe that it's possible to begin with. A lot of people just don't believe that, "Oh my gosh, we can do something different, and people will like that!" Because they look around and they say instead, "What do people like today?" That is called market research. And market research just looks at how people use a technology now. It doesn't pinpoint those people who have up till now avoided a strange and disruptive technology. Take the early smartphones, which promised the power of a computer in a mobile device. Market research didn't venture outside the circle of early-adopter technophiles to talk with potential users such as teachers and doctors.

Or take the personal computer. When we started Apple, market research only looked at how people used big, expensive computers that only companies could afford. Would companies want these little, puny, inexpensive microprocessors in the form they existed then? No. They couldn't do the big jobs. Market research didn't touch all the normal people making normal livings—dentists, teachers, doctors, lawyers, people in small businesses, and engineers. Apple rejected the near-sighted focus of market research on existing users, and focused its vision on potential markets of new users in the future—and that's what made it so successful.

You have to have a lot of confidence to take the leap past market research myopia. It's so hard to do something the first time. You'll creep and creep, and be so afraid to make the move. You aren't sure you can do it. You aren't sure it'll be successful. And then once you've done it and it's accepted, everyone jumps on and says, "Hey, this was so easy and obvious to all of us!" But it really wasn't.

Stern: Most people are risk-averse. They are afraid to make mistakes. As an inventor and creator, how do you accept making mistakes? And what motivates you to keep inventing and creating despite making mistakes?

Wozniak: As I listen to other people talk about their attitudes to risk, I'm convinced there are different types of people. Some of them just have a cold, calculated risk management system—like embedding a probabilistic formula in a spreadsheet. They expect to make mistakes. They have to insert a few more resources—and they cover them up.

At the other end of the spectrum are the sort of risks that real inventors take. I know a bunch of inventors in the Inventors Hall of Fame who have thought up some of the greatest things we have. These great inventions come almost by happenstance, by surprise, by serendipity. The people who invent them are very much like myself—more than any other group I've ever been in. They think independently, on their own in their own heads. They have ideas. They want to go a different way. They want to run into a laboratory, build something, test their ideas, and prove them. Their motivations aren't salaries, and stock options, and houses, and titles, and awards. Their motivation is really that they thought of something and they want to see if it's possible to bring their own little mental puzzles and games out of their heads and into the world at large.

Stern: Do you think there are certain skill sets that you need to have to be an inventor? Are you born with them, or are these things you're somehow taught over time?

Wozniak: There are so many different kinds of people, and a few are inventors. You could say, "Well, inventors were born that way." But we have no way to test that idea. Myself, I think we're all born equal and the accidents of life determine how each of us develops. What is your family like? Do they encourage you to explore, to be curious, to be creative? Do they put up barriers to exploration for the sake of "proper" conformist behavior, or do they leave the avenues to exploration wide open and smile and laugh when you do things that are even a little bit naughty? So it starts with parents in the home and accidental encounters with other kids you kind of like and enjoy. They become your best friends in school. Are they a little bit on the wilder or crazier side, and willing to explore? Also, the future inventor needs the good fortune to somehow acquire enough in the way academic skills to have access to learned knowledge about the world—which is not necessarily the same as thinking ability.

In my case, I had to understand something about how electrons flow in wires, and in chips, and in lightbulbs. I had to have some fundamental understandings in mathematics. But you don't really need that much of it, and it doesn't have to come from school. You can get your education in an awful lot of ways. I had no computer classes in school. Anytime I encountered a book that had anything about computers, I read, and I read, and I read. I studied and tried in my own head to put it together. Shyness actually helped me. I've heard other people talk about shyness leading to inventorship because—you know what?—if you're shy, you're not socializing like everyone else. You don't have to fit into the crowd. You can think your own thoughts and not be afraid of them. I was so shy, I was afraid of ever having to deal with people. I was afraid of conflict, of a discussion with people. I'm obviously not that shy now. But I was so scared.

In the months and years I was in the Homebrew Computer Club, I never once raised my hand. I'd be too scared to talk to someone else that way. But if I created great things and they came to talk to me—Wow! It was nice to have somebody talking about my things.

I just want to develop something for myself, not based on anything anyone else has ever done before in the world. It lets me be more of an independent thinker. I have some ideas in my mind that are different than the normal approaches that people are after. I don't want to go the normal way. I'll get trounced by everybody who knows how to manage teams, and hire engineers, and get resources, and do all that stuff. I would rather go off in the way that doesn't really have competition, doesn't have that much accountability yet, and just make and achieve something. I find that a lot of other inventors who get noted are sort of the same way.

Stern: Your mantra is "one step at a time." So how do you teach perspective or pondering?

Wozniak: I disagree that it's a mantra. The way a human mind organizes the world proceeds upward in a specific age-correlated sequence. Certain ages tend to reach certain mental stages that are testable. For example, your mind isn't really ready for concepts like algebra—a letter standing for something— until about seventh grade. The individual mind goes through these stages, and it's careful to go up that ladder a single step at a time. When you try to take people two steps up the cognitive development ladder—to bring a five-year-old child to the seven-year-old stage without passing through the six-year-old stage—they don't get it. It's important that we go through the intermediate steps. The development of my early understanding went one step at a time, closer and closer to my mature understanding of how atoms work, how electrons work, how little circuits can make logic, how logic can be combined to make things that are understandable, how to build little chips into counters, how to make tones in a phone, how to make television circuits. Later, I had to go up the exact same kind of ladder to make the Apple II computer.

Stern: But what makes some people ponder something or just want to think about something?

Wozniak: Well, you know, some people are probably born more of the dreaming type. But, like I said, I think it really comes from your environment and the early stages of your life. Did you have the time and environment to think things out a lot? Now, some people are ADHD: they're just reacting real quickly all the time. They don't have that quiet solitude to think things out on their own with nobody else around and with nothing to distract them from focusing on one thing—just white noise. You need quiet solitude to let your mind go in a direction it wants to go and think, think, think. And eventually you may pull out pen and paper and write some of your ideas and see if you have something that sort of works. It might be a poem. It might be a song. It might be an invention.

Stern: Do you have any particular tools that you like using for getting your ideas out there? Is it writing? Prototyping? The computer?

Wozniak: When I was young, I would go to sleep thinking about computer programs, or computer hardware problems, or math problems in school. Sleep and think and think as I fell asleep, working the variables in my head, seeing the lines of code. And then I would wake up in the middle of the night sometimes and have a solution that would save one line of code in a program. And so I believe very much that your mind is working while you're dreaming.

But I don't wake in the middle of the night with solutions very much. I'm usually so busy and so tired that I fall asleep really well. But in the morning and in the shower—my most quiet times—I always make up jokes. You know, coming up with your very own joke is as creative as making a computer—and once in a while they're really good. Or coming up with a thought for the day—a thought about society or humanity that you might apply and talk about, and then it becomes part of your inside thinking. So I think a lot, but my thinking has to be in the quiet.

Somebody might run up to me with a little problem and I just won't answer. And then while they're gone, I'm thinking, thinking in my head, and I come up with my ideas as to what the best approach to solving this problem is, who to call, what questions to ask. I like to reverberate the problem from different mental surfaces and think it out and ask the devil's-advocate questions and get to a deeper conclusion than people who answer reflexively right upfront.

Stern: Do you have definitions for the words "creative" and "innovative"? Is there a difference?

Wozniak: Oddly enough, no creative person wants to be definable. Creative people don't want other people to be able to predict what they're going to do. I don't have a strict definition of creativity. It involves difference from prior ways—coming up with ways that other people haven't thought of before. But creativity also involves value. I've run into artists who do very, very different things than anyone's ever done, but that doesn't mean that they have value. In our world, especially the world of technology, value is essential. The different way has got to come back around to something that people want to use and that its creator makes money with. Maybe it has many fewer parts, or maybe it's easier to use. Often it doesn't get recognized as creative and yet it might be.

I believe there are an awful lot of technical people who are just as creative as Steve Jobs/Steve Wozniak. They're out there and they create these things, but their creativity doesn't receive name recognition in a mass market, which is focused on the charisma of the big successful company. To a creative person, that doesn't matter. To a creative person, it just matters that, in your head, you know you are creative and you've done something exceptional that other people haven't done. And you cherish that knowledge, because you never know if you're going to lose your creativity the next day and never have it again.

Stern: Is innovation any different than creativity?

Wozniak: I think of those words as meaning very much the same thing to society and most people today. But innovation can have the more narrow meaning of making small improvements to existing products. Creativity, on the other hand, suggests conjuring something big and unexpected out of thin air and dropping it on us. A whole different approach or a whole different thing in life that we never imagined having before and that's usually disruptive. It causes businesses to fail, like RIM and Nokia. It causes big huge companies to just totally go out of business because it totally disrupts what they had before. Maybe the traditional mass media industries feel a bit of that.

Stern: Do you think innovation is more on the commercialization side of new ideas?

Wozniak: It's so hard to say, because I think any person who does something creative would also call himself or herself an innovator. Does the idea add one feature to a computer that improves its speed a little? Incremental changes can be just as creative as disruptive changes when they come from the same impulse to make things better in a new way—rather than just engineering by the book, exactly as I learned it in school, exactly as I've done it before, the same thing over and over. That's why it's so hard to define a particular invention as either creative or innovative, and not both.

Stern: You invent by yourself. In corporate settings with all the technology and the complexities of everything today, inventions are produced by groups of people. Do you think that an individual can be inventive within a team of people?

Wozniak: I absolutely believe that. Now, at Apple, after we got larger and had teams of people, somehow I was different. I was the founder. I could just go off and work on my own projects outside the management structure of having to write forty-page reports on every little episode and what every tiny little shortened routine in my code program is going to do. I can't operate that way. I want to just go at it on paper, flow my ideas out, start building some chips, write a little code, get another idea, add some more code to it. I just want to go with the flow without impeding the intuitive thought that's so important.

Stern: Along the way, have you had mentors? And do you mentor people now?

Wozniak: I didn't really have direct mentors apart from my father. Although he was an electrical engineer, he didn't lead me towards electronics. I went there on my own initiative out of interest. He would get out a blackboard and teach me things about charges and particles, how they go through transistors, and what the formulas for transistor amplification are. He would teach me this kind of stuff that I'd never learn in school. Without him, I wouldn't have climbed any of the little steps that were so big to me in their day, and which led up to the objectively big steps I would take as an inventor.

I also had a high school teacher who taught electronics. He had a better course and more equipment than any of the local colleges. He knew where our minds were attracted by the current developments in electronics, and he always approached the class and subject material from that point of view. He had come from the military. Some of the great instructors I've had came from the military—as do many great mentors and leaders. So this high school teacher had strict procedures and he had stockrooms full of parts that he had gotten companies to donate. He spent a lot of his own time to make his class great. He contacted companies and found engineers who would let students like myself go in and work on projects such as programming a computer to control streetlights.

When Steve Jobs and I were first starting Apple, we went to the world's first computer trade show out in Atlantic City, called PC '76. We were just introducing the Apple I, even though we had the Apple II designed and working. There were a bunch of little companies with two young kids like us—with no money and ideas that maybe these microprocessors could make the start of a business. It was the garage-style entrepreneurship that you see nowadays with kids going into writing apps for iPhones and the like. So we were talking to a couple of young people at the show, and they had a metal case around their device. We asked them how they made it. Oh, they'd had help from an older man who went around giving help to young people when they needed it on their projects. He called himself a "mentor." And ever since then, I've just held that thought so dear to my heart. It's so important for young people starting a company who don't have business experience or money to get that little assistance that lets them go so much further.

Stern: Do you do that with young people today, then?

Wozniak: I do it every chance I get. Now, I don't have a lot of time to be going out and manufacturing metal parts or even wiring the circuits myself, but I sure have time to talk and listen to young people's ideas, try to point out a few little deficiencies, and suggest places they might go where they will find answers. Now there are so many answers on the Internet that they don't necessarily need mentors for all their questions. But I'm willing to give them inspiration and encouragement: "You can do it. Go as far as you can on your own."

Stern: Where do you seek inspiration?

Wozniak: Being in the technology field, I'm around a lot of technologists, and they all have ideas: "You might do this. You might do that." I might be out bowling or playing Segway polo, and somebody will toss out a little idea when I bring up a subject that's on my mind. I talk openly about little projects that are interesting to me, whether or not I'm actually constructing them, as I did in the old days.

Stern: Do you have any professional heroes?

Wozniak: You know, I do. I really admire Bill Gates. I don't think he was the visionary, or great technician, or company and product thinker that Steve Jobs, and even myself were, but he has a good heart and he does a lot of philanthropic work now. He's just a good person, and that's much more important than being the richest man in the world. Oddly enough, I barely ever met him, even though he was one of the young people at PC '76.

Stern: How do you start a project? Do you have a particular ideation process that you go through?

Wozniak: I'm going to go back in time. I could build things and be the technician, but before I'd start wiring something together, I'd have a clear picture in my head. I would see the goal—what it was going to be. I didn't even consider what a computer had to do for other people. I just knew that it had to do what I needed it to do for me. And that was my ideation process, right up to the Apple II.

After I had an idea for a project, I normally started it off by being aware of the building blocks. It's like if you were going to design a house, you'd need to know all of the materials that are available. I would see in my head where my chips should start going, but I would always reduce the architecture and design it around my actual building pieces, which were constrained by their affordability. I was very much into conserving and saving then. It doesn't matter nowadays, when we have a billion parts on one little chip that costs twenty-five cents and we don't have to worry if anybody sees it.

Stern: It sounds like you try to approach problems from a materials and processes understanding?

Wozniak: Materials and processes are important, but I approach them as means to reach a goal. Once a goal's in my head, it doesn't take very long for me to sit down and get to wiring pieces together to do what I had envisioned.

Stern: Do you have a next project you're working on that you want to talk about?

Wozniak: I don't have a specific project right at this moment. I would sure love to investigate building full CPUs [central processing units] out of GPUs [graphics processing units] combined with NAND flash memory and some DRAM in the right proportions to get ultra-performance. GPUs take an awful lot of power because you've got so much computing all at once, so I would love to develop chips that work on photonics, since electrons are so much lighter and more efficient than atoms. To make a chip with photons, electrons have to be made to go through little pipes and do the right things and pop out little answers, and I know how to make them do that. The diode, which lets electrons go through one way but not the other, was the heart of all the little logic things that I built when I was young. So I'd like to develop a logic gate for GPUs.

Stern: Do you have any advice for the would-be inventors out there?

Wozniak: Examine your life, I would say. Whether you're in elementary school, middle school, high school, or college, you have ideas in your head for the things that you would like to have in your life that don't exist. As a would-be inventor, you will find yourself much more driven by your personal ideas than by external factors. Wouldn't it be nice to have a machine out on your driveway that washes your car all night, going slowly, one square centimeter at a time? When you wake up, your car is beautifully and effortlessly washed. But it's not going to happen unless some sixteen-year-old gets in his mind, "I want to make this my project." It ends up as his master's project in college, or he just goes out and makes the machine on his own—who knows?

You'll have another iRobot, Roomba, or some other successful product out there. That's how a lot of these things really get done. It's a drive from the personal human side and not influenced by salary, stock options, and all that. If you're really motivated by those external markers of success, that's the sort of person you're going to turn out to be, probably.

Stern: Apart from Roomba, electronics, and Apple products—do you have a favorite invention that you like to have around with you all day or in the home?

Wozniak: My early old-school Game Boy with *Tetris* playing on it.

Stern: That's electronic—so, unrelated to anything electronic…

Wozniak I would say my Segway. It's so unusual. You look at it—who could have imagined this *Jetsons*-like device? You stand on it. That thing takes so little space to park anywhere. Inside of a restaurant. I ride it down to town, and I'm smiling. I enjoy the wind going past me. It's almost like I'm partly skiing into town. As a matter of fact, I carry it in my car and I explore new cities like nobody can do on foot. And I play Segway polo. So that thing is really a great little device in my life.

If I thought further, I'd say diet cherry Dr. Pepper. Music is hugely important to me—not any particular genre. I love to go to concerts. I go to maybe fifty concerts a year. I drive long distances to go to the little tiny music bars where people are listening to the words, and hugging themselves, and falling in love, and the music is a magic dust. The important thing is find what makes you happy. Keep a lot of happiness in your life. I love to just make jokes wherever I go.

Stern: Any final word of advice for would-be inventors?

Wozniak: First of all, believe that your idea is possible to do. If you assume that it's not possible, you'll just give up early. Don't take money real quick. Don't start a company by a formula. Have a job, get some income, work on your own on this dream that you want to change the world on. Don't worry that others might beat you to it. Take the time to think it out and do it perfectly. Have something really good to show before you raise money, and you'll own a lot more of what you deserve in the end.

Jobs was known as a micromanager. Those little tiny details and visual cues make so much difference. What people see has to be like real life. If you're going to be in the technology field, make your technology seem more like a human. Look at what you're doing and what other people are doing. Yes, you'll probably go through some hard money times on the way. But I wouldn't say, "Give up all your income betting on your big idea."

Stern: You said, "Believe in what you're doing." How do you know if something's not working or when to give up?

Wozniak: When I did all the stuff leading up to Apple, I thought many of my ideas weren't going to work. But I went ahead blindly anyway and said, "I'll solve that problem when I get to it." As it happened, every one of them actually wound up working. In later life, yes, I've had projects fail, where I could not achieve my technical goals. That's one of the good reasons to start a company with as much of a working model as you can, to really, really be sure that it is workable. That's the main thing that'll set you back: finding after you start up that your idea is not workable.

Also, don't assume that just because it's a good product—that people need in their lives and will help them—that it will necessarily sell. Deciding how to get it out and market it are so important. The best strategy might be to penetrate a little at first, and then a little more, and a little more until it gets well accepted. You might have to go search around with marketing experts, figuring out, "Where do we sell this product?" You can't just put it into Best Buy—that's not how the world works. You've got to look for the right kind of outlets and progressively build your business up toward the level of the big companies with the big products. Don't expect to be there on Day One with a start-up.

Stern: What have you learned from all this so far?

Wozniak: Be good. Learn. Try to know more than other people, even if it doesn't apply to your grades. Pick up books. Read them because you're interested in them, even if you're not taking the class. Learn stuff. File it away. It's not useless. You might feel like you're on your own and different from other people. Don't let that bother you. The important thing is that you know that you're making yourself happy.

Steve Jobs didn't finish college and had no money in the bank at all and no savings accounts, no relatives or friends who could loan us money, and no business experience. We had never worked in business, in a company. We'd never taken a business class. I don't think we'd read any business books—or, if so, it wasn't me. We just reverse-engineered the common sense of the consumer. In the end, the end user is the one to focus on. You have to relate to them very well.

Stern: Do you plan to retire? And what will you do?

Wozniak: Well, I always keep dreaming about the day when all I'm doing is fishing all day long. I don't fish, but maybe I'll just be playing my Game Boy and computer games. But life got so busy, especially with the Internet. In my situation, where I have thousands and thousands of friends on Facebook, and I don't know them but I'm constantly interacting with them and answering their questions. It takes a lot of my time. I don't know if I'm going to get to retire.

Index

G, H, I

Made in the USA
San Bernardino, CA
18 July 2013